Potential Therapeutic Applications of Nano-antioxidants

Sharda Sundaram Sanjay • Ashutosh Kumar Shukla

Potential Therapeutic Applications of Nano-antioxidants

Sharda Sundaram Sanjay
Department of Chemistry
Ewing Christian College
Prayagraj, Uttar Pradesh, India

Ashutosh Kumar Shukla
Department of Physics
Ewing Christian College
Prayagraj, Uttar Pradesh, India

ISBN 978-981-16-1142-1 ISBN 978-981-16-1143-8 (eBook)
https://doi.org/10.1007/978-981-16-1143-8

This Springer imprint is published by the registered company Springer Nature Singapore Pte Ltd.
The registered company address is: 152 Beach Road, #21-01/04 Gateway East, Singapore 189721, Singapore

Preface

Traditional antioxidant therapies have been less effective in preventing various diseases caused by oxidative stress. Nanoparticle antioxidants have emerged as a new expansion of antioxidant therapies for the prevention and treatment of such diseases. It is our pleasure to present this book with specific goal of connecting the available literature to the therapeutic functions of nano-antioxidants. Chapter 1 presents the basics of free radicals and acquaints with antioxidants and their classification. Chapter 2 talks about the chemistry of phytochemicals as natural super-antioxidants. Chapter 3 presents an overview of nano-antioxidants and different methods of determination of antioxidant activity. Chapter 4 deals with the mechanism of antioxidant activities. Chapter 5 presents quantitative structure-property/activity relationships (QSPRs/QSARs).

We sincerely thank Dr. Naren Aggarwal, Editorial Director—Books, Asia, Medicine and Life Sciences, Springer, and Mei Hann Lee, Editor, Life Sciences, Springer, for giving us an opportunity to present this book. We also thank Priya Shankar, Project Coordinator (Books), for her support during the different stages of the publication process. We have tried our best to be simple in presentation style. We hope that you will enjoy reading the text and gain insights about the therapeutic applications of nano-antioxidants.

Prayagraj, India
January 2021

Sharda Sundaram Sanjay
Ashutosh Kumar Shukla

Preface

Contents

Chapter 1
Free Radicals Versus Antioxidants

Abstract Antioxidants are the compounds that prevent other molecules from oxidation. During the course of an oxidation reaction, free radicals may be produced. "Antioxidants" prevent the oxidation of species to check the production of free radicals. A free radical is a molecule lacking one electron to get stability, due to which it becomes highly reactive. The electrons tend to pair up to attain stability. Therefore, the generated free radical spins erratically, and collides with many other atoms and molecules in an effort to snatch electrons. In our body, numerous chemical reactions such as digestion and respiration occur to produce energy to stay us alive. As side effects of these biological reactions, our body system also produces free radicals as by-products. Free radicals are formed when biomolecules such as proteins, lipids, and nucleic acids are exposed to them, through redox reactions in the system itself and cause injuries to the organs/tissues/cells by damaging cell wall and cell contents, which in turn becomes the cause of serious ailments in the body. There are number of methods through which free radicals are generated. Antioxidants protect cells from damage caused by free radicals by regulating ROS-related enzymes and maintain health.

Keywords Free radicals · Antioxidants · Biomolecules · Enzymes · Reactive Oxygen Species (ROS) · Carotenoids

1 Introduction

The word antioxidant generally refers to a compound that prevents oxygen from being consumed. Antioxidants are compounds that prevent other molecules from oxidation. The loss of electrons from a material is referred to as oxidation and that electron is taken by some other species. In industrial processes, for example, metal corrosion prevention, rubber vulcanization, polymerization, and internal combustion engine fouling, antioxidants play a very important role. In the field of live sciences, majority of the antioxidants based researches centered on developing a method on how they can avoid unsaturated fats and lipids from oxidation that causing rancidity. Throughout the whole process of an oxidation reaction, free radicals may be

S. Sundaram Sanjay, A. K. Shukla, *Potential Therapeutic Applications of Nano-antioxidants*, https://doi.org/10.1007/978-981-16-1143-8_1

produced. As rising of sun removes the darkness of night, in the same way "antioxidants" prevent the oxidation of a species to check the production of free radicals. Antioxidants are discussed in detail in consecutive paragraphs, but to know the working of antioxidants, it is necessary to know about the origin of the free radicals. Let us discuss free radicals first.

2 Free Radicals

2.1 Formation and Definition of Free Radicals

We all know that an atom is the smallest unit of a substance. Each atom possesses electrons, which are negatively charged particles that revolve about a central positively charged nuclear core in specified orbits, to balance the nuclear charge. Two or more atoms combine together to form a molecule. If a molecule is formed by the sharing of electrons between the combining atoms, then the bond between them is known as a covalent chemical bond, which holds the atoms together in a molecule. But in some cases, as result of any chemical reaction, or due to some other intention, an electron is taken off from an atom of a molecule, then the molecule with a free unpaired electron is now called as a free radical, which is highly unstable (Fig. 1.1). Thus a free radical, on the other hand, is a molecule lacking one electron to get stability, due to which it becomes highly reactive. The electrons tend to pair up to attain stability. Therefore, the generated free radical spins wildly, colliding with other molecules to steal electrons. Thus free radicals are the species with an unpaired electron revolving around the nucleus, may be an atom, ion, or a molecule [1]. These tend to get stability, by extracting an electron from the nearest neighbor molecule. In terms of the human body, free radicals are described as: "unstable oxygen molecules that can do damage to our cells, organs, and/or body as a whole."

In order to understand how free radical causes damage (also known as oxidative damage) to our body, it is better to know first that what oxidation really means. The

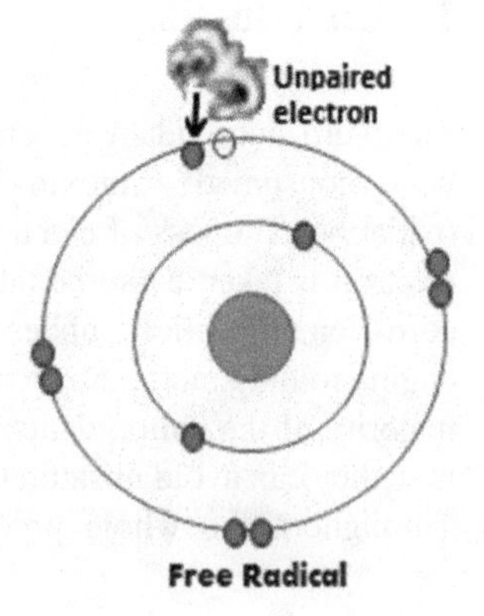

Fig. 1.1 An unpaired electron in a free radical

conventional approach to define oxidation in terms of electron transfer is, "Oxidation is a process in which loss of electron or electron density occurs" and its reverse, "reduction is the gain of electron or increase of electron density". From this explanation, it is easy to see why oxidation and reduction are inseparable processes. They occur simultaneously. Therefore these processes or reactions are known as redox reactions. In redox reactions, one substance gains electron (get reduced) and another one loses electron (get oxidized) to it. Thus a redox process operates as a driving force causing oxidative damage to the body tissues.

In our body, numerous chemical reactions such as digestion and respiration occur to produce energy to stay us alive. As side effects of these biological reactions, our body system also produces free radicals as by-products, which may be atoms or molecules with unpaired electrons. The unpaired electron may occupy an atomic or molecular orbital (Fig. 1.1). The presence of unpaired electron makes free radicals an extremely reactive species having very low stability. So as short-lived particles, these free radicals tend to attack neighboring molecules in order to re-establish a stable state and structure. To fulfil the purpose, proteins, lipids, and nucleic acids are among the biological compounds that react with free radicals within our body through redox reactions inherent in the system itself. This act of free radicals causes injuries to the organs/tissues/cells by damaging cell wall and cell contents, which in turn becomes the source of serious ailments in the body. The most bounteous free radical is oxygen because oxygen molecule itself bears two unpaired electrons. Paradoxically it is very true that a human body cannot survive without oxygen. To perform its physiological activity, oxygen is required in abundance. Therefore intake of oxygen cannot be avoided and at the same time, it really wears down the body's immune system, weakens structures, causes illness, and accelerates the aging process.

2.2 How Free Radicals Are Destructive

For the first time Dr. Denham Harman in 1954 observed free radicals were the cause of early aging and a number of chronic diseases while studying human biology and its interaction with ionization radiation. Sunburn is a remarkable example of free radicals in action. When sun rays comes in contact with the skin, the radiations cause mild "transfigurations," meaning creating some alterations in the texture of the skin and damaging it by producing more free radicals [2]. In aerobic species, mitochondria act as the primary source of energy to perform the cell's vital functions. To meet out this requirement, it produces ATP through chain of redox reactions. With the electron transport system, ATP binds to the tricarboxylic acid cycle. That happens as a result of the oxidative breakdown of food and the synthesis of NADH and FADH2 in different metabolic routes. Glycolysis, β-oxidation, and the Kreb's cycle are all examples of metabolic processes. These reactions, on the other hand, are inevitably linked to the production of unstable oxidative compounds (ROS) that have unpaired electrons. They recapture electrons from neighbor biomolecules to stabilize

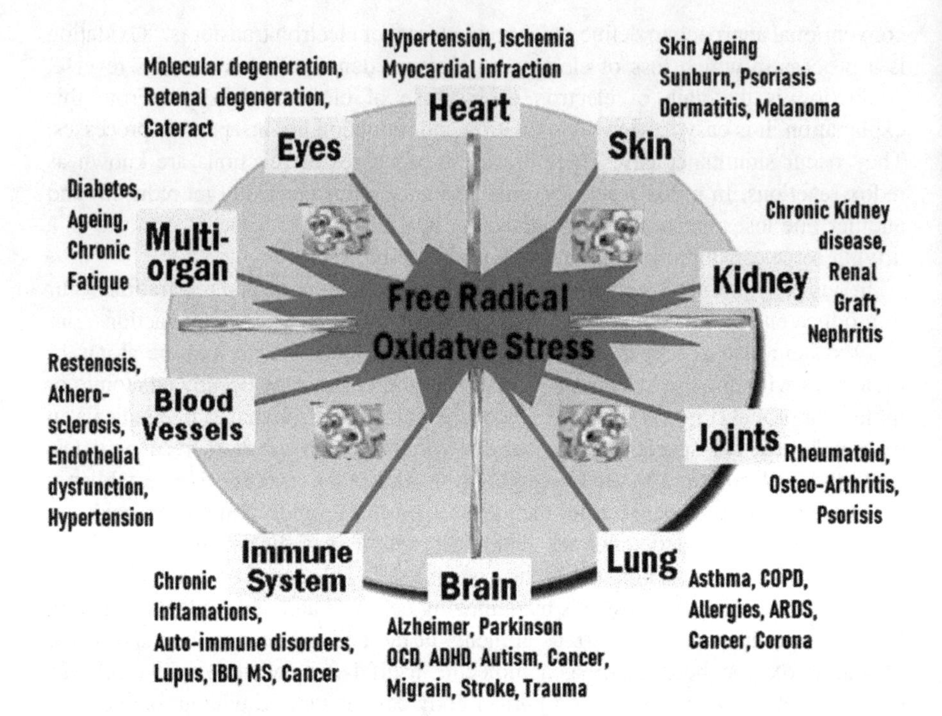

Fig. 1.2 Damages and diseases caused by free radicals

themselves, and making other destabilized. As a result, they are unable to carry out their responsibilities effectively, causing homeostasis to fluctuate and eventually as a consequence cellular damage occurs. Organisms are not only affected by the imbalances in the redox reactions of their metabolic reactions, but the atmosphere on our planet also contributes as having its own existing oxidant characteristics. Moreover, radiation, pollution, microbial pathogens, xenobiotic toxicants, dietary contaminants, and many other exterior influences also play their cumulative role in increasing free radical concentration. Establishing appropriate mechanisms to reduce oxygen poisoning and the disparities in redox reactions is viewed as a significant challenge among scientific community.

Damages to our cells may cause damages to our organs, and damages of our organs in turn may lead to Alzheimer's disease, cancer, heart disease, inflammatory, and other serious health problems. Free radicals possibly attack various proteins including DNA and might even be involved in the emergence of cataracts and the aging process and many more (Fig. 1.2). Now, the question arises that if it creates such a havoc then why our body produces such a dangerous entity?

2.3 Production of Reactive Organic Species

First free radical inside the body, the superoxide free radical, was discovered around 1970s. Since that discovery, numerous free radicals have been discovered. A very popular and hazardous category of free radical is something that involves oxygen.

The term for all this is "Reactive Oxygen Species" (ROS), an oxidant entity. Some of the ROS are as shown in Table 1.1, below:

All the mentioned ROS are competent enough of causing cellular damages by interacting with membrane lipids, nucleic acids, proteins, enzymes, and many other smaller biomolecules. Actually reactive oxygen species (ROS) are produced through several mechanisms within our body and the majority of free radicals formed by cells are by-products or upshots of:

- *as a result of regular aerobic metabolism:* The mitochondrial electron transporting process gobbles up about 90% of the oxygen being used by cell and ROS are produced as by-products.
- *the phagocytes' oxidative explosion (act of leucocytes, the WBC):* As part of the denaturation mechanism of foreign proteins (antigens) and the killing of bacteria and viruses by WBC, ROS executes their prime role.
- *xenobiotic metabolism,* i.e., detoxification of toxic substances. ROS here helps in detoxification.

As a result, factors such as vigorous physical activity, which intensifies metabolic processes, chronic inflammations, infectious diseases, and other ailments, vulnerability to toxins, the risk of "leaky gut" disorders, drug or toxin consumption, such as tobacco smoke, pollutants, contaminants, pesticides and herbicides, can all lead to a rise in the body's natural oxidative stress contents. The second most common free radical source is unsaturated molecules, i.e., molecules having multiple bonds. The multiple bonds of unsaturated molecules being unstable make the molecules very prone to transform into free radicals. Unsaturated fatty acids are perhaps one of the most important unsaturated biomolecules of our body, as a vital constituent of the cell membranes or cell wall. The inner cell organelles are protected from the external vulnerability by the cell membrane. The oxidative breakdown of fatty acids in cell membranes could result in additional free radicals and chemicals that may pass through the cell membrane. This intrusion continues to the nucleus of the cell and due to which, our genes and DNA also come in threat. Genetic changes may occur if DNA is affected and could ultimately lead to kill the host.

Table 1.1 Some reactive oxygen species (ROS)

Free radicals	Particals, which are not free radicals
Superoxide, O_2^{-}	Hydrogen peroxide, H_2O_2 (Fenton's reaction)
Hydroxyl, $OH^{\cdot}$	Hypochlorous acid, HClO
Peroxyl, ROO ·	Ozone, O_3
Alkoxyl, $RO^{\cdot}$	Singlet oxygen, $^1O_2$
Hydroperoxyl, $HO_2^{\cdot}$	Nitric oxide, NO

Other than natural sources, there are number of ways, through which free radicals are generated. These include:

1. Thermal Cracking: When a substance is heated above 500°C without oxygen, thermal cracking occurs. It is mostly used to dissociate alkanes having high molecular weight into smaller alkane and alkene remains.
2. Homolysis of Peroxides and Azo Compounds: Thermal susceptible peroxides (-O=O-) and organic azo compounds (R–N=N–R) can dissociate into peroxy, alkyl, and nitrogen radicals.
3. Photolytic Bond Homolysis: In the presence of sun light,organic halogen compounds can easily form free radicals.
4. Electron Transfer: Electrons from the oxidation of iron (II) ion in the presence of peroxide can be considered as radical initiators.
5. Hydrogen and Halogen Atom Abstraction: In the presence of oxygen, alkanes can form radicals that will in turn form peroxides if the radical is left unused.

2.4 *Not Always Damaging*

Free radicals are actually unstable molecules that are attempting to recover their equilibrium. If this is really the case, then issue of why and how free radicals thrive raises. Free radicals come to an existence because of the number of reasons. They may be produced by any of the two ways, one that can be controlled and the other one that cannot be. The reactions involved in the metabolic process are autoregulatory. We cannot control such reactions. But it does not mean that they always cause devastation. They serve useful purposes too for us. Complete elimination of these free radicals from the body is not possible, and their complete elimination is also very dangerous too.

Our body needs some free radicals for normal functionalization. They are required, for example, to produce melanin that imparts color to the skin. When we respire or ingest food, we consume some energy to initiate some chemical reactions or metabolism to be done. As a side effect of these reactions, free radicals are formed naturally. These reactions are autoregulatory. Our immune system is yet another interior unmanageable source of additional free radical generation. Our internal security guard creates free radicals to fight off and neutralize bacteria, viruses, and harmful diseases when required, i.e., generation of free radicals occurs as a result of defense mechanism.

2.5 *When it Is Alarming?*

As we have discussed, some free radicals that are naturally produced within our body, as natural by-products, are generally manageable. It only becomes a concern whenever the level of free radicals generated surpasses our body's capability to

Table 1.2 List of some of the environmental factors that cause the production of free radicals

• Stress(emotional and physical) generates toxins	• Environmental & Industrial Chemicals	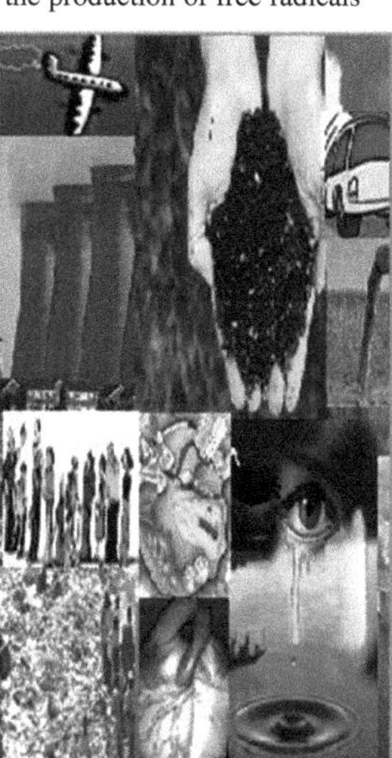
• Pollution	• Synthetic materials	
• X-rays Radiation (which includes UV rays from the sun)	• Household cleaners	
• Radioactivity	• Medications	
• Food additives & Processed foods	• Drugs	
• Ozone depletion	• Recreational and prescription	
• Airline travel	• Pesticides	
	• Herbicides	

handle them. This reflects as a serious disorder resulting in the austere damages to our body. Here it is important to mention that oxygen may not be the only free radical to be focused about, but this is the most highly prevalent. By definition, any unbalanced molecule can act as a free radical.

Free radicals that produced biologically within the body keep us alive. But there are numerous environmental factors which enhance the free radical production greatly causing disturbance in the handling capacity equilibrium and are proven to be more dangerous. These environmental factors may be tobacco smoke, air and water pollutants, radiations, insecticides, fungicides, herbicides, toxins in our food supply, vehicle emissions, and a variety of other factors all contributing to our health. The following Table 1.2 lists some of the various external damaging factors and sources that enhance free radical production.

Although it's not really an exhaustive list, but it illustrates how environmental factors play a role.

3 Antioxidants

Now, here comes the antidote of free radicals combating against it, the antioxidants.

3.1 Definition

As oxygen is essential for our survival, nature has provided it in abundance. On the other hand, nature has also provided us with oxidation antidotes to fight against oxidants for survival, subsequently called as antioxidants. Fortunately, various beneficial antioxidant compounds control the formation of free radical naturally. The term "antioxidant" may be defined in number of ways:

- *An explanation in chemical terms:* "a substance that opposes oxidation or inhibits reactions promoted by oxygen or peroxides".
- *An explanation from a biological standpoint:* "synthetic or natural substances that prevent or delay deterioration of a product, or are capable of counteracting the damaging effects of oxidation in animal tissues".
- *Definition given by Institute of Medicine:* "a material that greatly reduces the negative effects of reactive species like ROS or RNS on normal human physiological functions" [3].

Antioxidants fight free radicals to keep cells healthy and resist cell damage by reducing the incidence of chronic diseases through preventing cellular degradation caused by oxidative stress. Ginseng, for example, produces ginsenosides, which are stimulant compounds with antioxidant properties that protect the vascular endothelium from damages caused by free radicals [4].

Antioxidants also play a key role in regulating the release of ROS-related metabolites. Antioxidants effectively inhibit the functions or signals of free radical producing enzymes including NADPH oxidase and xanthine oxidase, which produce free radicals in the cell (XO) or just by boosting both activity and expression of antioxidant enzymes like superoxide dismutase (SOD), catalase (CAT) as well as glutathione peroxidase (GPX) [5, 6]. The low availability of antioxidants causes profuse cell damages that may become snowballing and devastating.

3.2 The Locations where Free Radicals and Antioxidants Counteract

Every cell needs oxygen to generate energy through metabolism, and thus producing unstable molecules, and creates "free radicals." Antioxidants check, resist and counterbalance free radical damage and free radicals formation. It means that each and every cell is an arena for the activity of free radicals and the antioxidants' counterattack. Our bodies have developed a powerful sophisticated and advanced antioxidant defense mechanism to defend our cells and organ systems from reactive oxygen species. It comprises a number of constituents, both extracellular and intracellular in origin, that interact and work collectively together to counterweight free radicals. The following are some of the elements of a complex antioxidant defense system:

- Vitamin C, tocopherols and tocotrienols (vitamin E), carotenoids, as well as other low weight molecules including glutathione and lipoic acid are sources of nutrient-derived antioxidants.
- Antioxidant enzymes that stimulate free radical passivating interactions, such as superoxide dismutase (SOD), glutathione peroxidase (GTP), and glutathione reductase (GTR).
- Ferritin, lactoferrin, albumin, and ceruloplasmin are metal binding proteins which repeal and replace free iron and copper ions, which can catalyze redox degradation reactions.
- A number of other antioxidant phytonutrients present in large and diverse group of plant sources.

Each antioxidant mentioned above in different categories plays a unique and important role in assisting our body's requirement in their fight against free radicals. A person's body does seem to have a number of mechanisms operating to combat the damages caused by free radicals and also other reactive oxygen species (ROS). For each free radical which campaigns against us, there are antioxidants in our favor to counteract those campaigns. As a consequence, antioxidants are an essential line of defense against damages caused by free radicals.

3.3 Classification of Antioxidants

Antioxidants can be classified on the basis of its different attributes:

3.3.1 Classification on the Basis of the Functional Attributes

In this category, antioxidants may be classified as primary and secondary:

1. First-line Primary antioxidants:

Such antioxidants encompass chain-breaking moieties that respond with lipid radicals to transform them to much more stable compounds. The majority of primary antioxidants have a phenolic component, which include antioxidant minerals, antioxidant vitamins, as well as phytochemicals such as flavonoids, catechins, carotenoids, −carotene, lycopene, diterpene, black pepper, thyme, garlic, cumin, along with their variants [7]. A detail description of these are given in subsequent chapter.

2. Secondary antioxidants:

Secondary antioxidants include phenolic combinations which trap free radicals and terminate chain reactions from continuing. The following are some of the examples:

(a) Butylated hydroxyanisole (BHA):

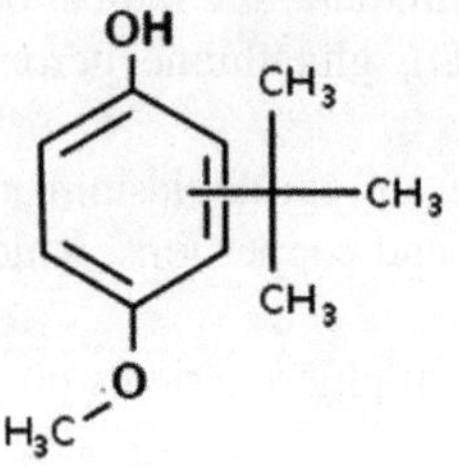

It is a mixture of two isomeric compounds. Since tertiary butyl group is juxta positioned to the hydroxyl group, it is also known as “hindered phenol” as it may hinder its efficacy in vegetable oils but enhances its carrying potency.

(b) Butylated hydroxyl toluene (BHT):

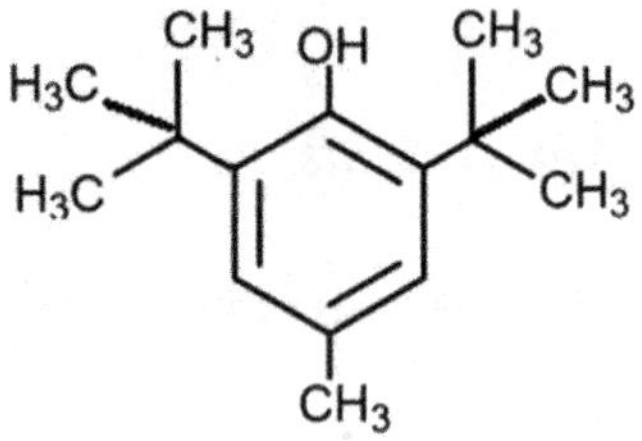

It is a sterically hindered phenol. It is very much susceptible to loss through volatilization at high temperature applications. Generally it is used with BHA and citric acid/BHT combinations.

(c) Propyl gallate (PG):

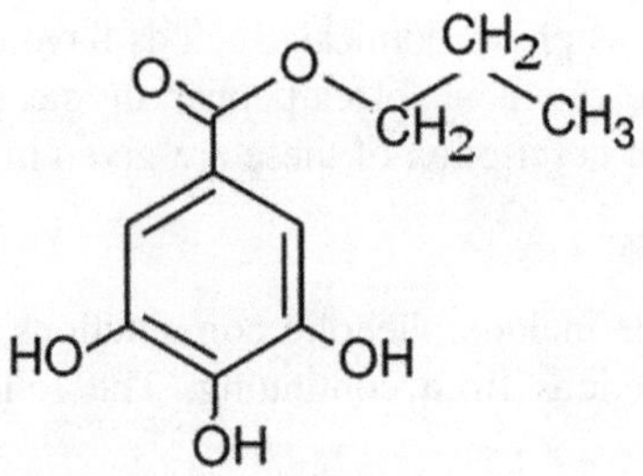

Due to the presence of three hydroxyl groups, PG becomes very reactive. It is having lower solubility, tending to form chelated colored complexes with trace minerals such as iron. In alkaline medium, they are heat labile. It may be used solely or with BHA or citric acid/PG combinations.

(d) Tertiary-butylated hydroquinone (TBHQ):

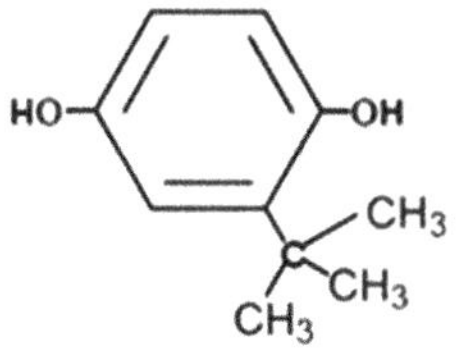

It is exceptionally very efficient antioxidant. It has been mostly used for non-edible applications before gaining approval in any particular food material.

3.3.2 Classification on the Basis of Enzymatic and Non-enzymatic Attributes

In this category, the antioxidants in living cells are divided into two categories: enzymatic and even non-enzymatic antioxidants.

The above division can be further segmented into subcategories. Antioxidants that are released by enzymes, i.e., enzymatic antioxidants, participate in enzymatic reactions, both primary and secondary enzyme based defenses within our body [8].

Three essential enzymes that inhibit the production through inactivation of free radicals are in control of the primary defensive lines. These are: (1) glutathione peroxidase, which forms selenols by sacrificing two electrons to eradicate peroxides through reduction. Peroxides are therefore ruled out again as alternative Fenton reaction precursors. (2) Catalase is a potent antioxidant that is currently recognized. It turns hydrogen peroxide into water and molecular oxygen. For about 6 billion hydrogen peroxide, only one molecule of catalase is sufficient, and (3) superoxide dismutase is just an enzyme that converts superoxide anions to hydrogen peroxide as a precursor for catalase action [9].

With glutathione reductase and glucose-6-phosphate dehydrogenase, the secondary enzymatic defense manages its front. Glutathione reductase converts glutathione (an antioxidant) from its oxidized to reduced state, annihilating further free radicals throughout the process [10]. These two enzymes work in tandem with the primary antioxidants in the enzymatic defense system and do not specifically neutralize free radicals. Non-enzymatic antioxidants are divided into many categories, the most important of which are vitamins (A, E, C), enzyme cofactors (Q10), minerals zinc and selenium, peptides (glutathione), phenolic acids, and nitrogen compounds.

3.3.3 Classification on the Basis of the Occurrence of the Antioxidants

There are certain antioxidants, that are produced during the course of the body's natural metabolism and certain are taken through dietary intake that contains number of smaller antioxidant molecules. Thus, on the basis of the occurrence of the antioxidants produced within our body or outside the body, they are categorized as:

1. *Endogenous:* There are several scavenger antioxidant enzymes produced within the body itself to combat against free radicals. Free radicals are gobbled up by these endogenous enzymes, which transform them into benign molecules.
2. *Exogenous:* Food and plants are external resource of micronutrients that serves as antioxidants. Being external resources, these are called exogenous antioxidants.

There are several defense mechanisms occurring within our body endogenously to facilitate and defend organs and tissues from cell disruptions caused by free radicals. There are certain antioxidant enzymes that require some transition metal ions as micronutrient cofactors for optimum catalytic activity. Glutathione peroxidase, catalase, and superoxide dismutase (SOD) are such enzymes which metabolize oxidative hazardous adducts to benign ones. Superoxide along with hydroxyl radical is two significant culprits. To combat the superoxides, endogenous enzyme SOD entrenches, which require zinc, manganese, and copper ions as cofactors. Generally as a consequence of poor eating habits and inadequate dietary intake of these trace minerals, only a minimal amount of SOD is formed, that compromises the efficiency of the antioxidant's defense mechanism. With the increasing age, the consumption and absorption of such essential trace ions/minerals generally decreases.

Another most destructive free radical known to biochemistry is the hydroxyl radical. It is the direct causative of cancer. The body counteracts it with endogenous glutathione peroxidase enzyme (Scheme 1.1). The amino acids glycine, glutamate, and cysteine come together to form a water-soluble antioxidant glutathione. Glutathione peroxidase, an enzyme derived from it, efficiently inhibits the conversion of

Scheme 1.1 Glutathione: a tri-peptide

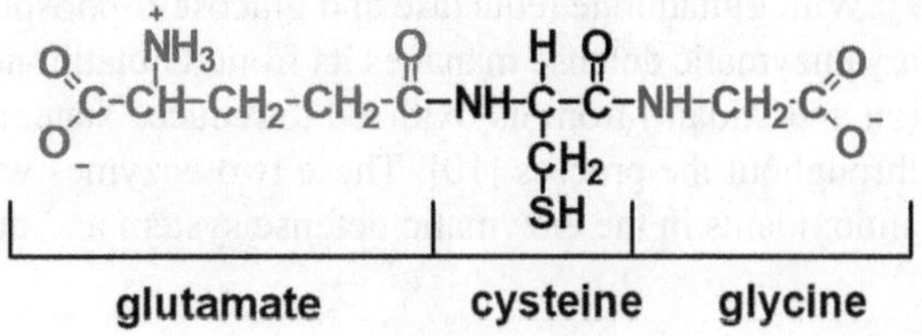

$^{-}O(O)C\text{-}CH(NH_3^{+})\text{-}CH_2\text{-}CH_2\text{-}C(O)\text{-}NH\text{-}CH(CH_2SH)\text{-}C(O)\text{-}NH\text{-}CH_2\text{-}C(O)O^{-}$

glutamate cysteine glycine

Table 1.3 Location and function of some antioxidants in mammalian/human cells

Antioxidant Component	Nature	Finding location	Nutrient mineral/ compound involved	Functions allocated
Superoxide dismutase	Enzyme	Cytosol	Cu, Zn, Mn	Conversion of superoxide to hydrogen peroxide
Glutathione peroxide	Enzyme	Cytosol	Se	Conversion of hydrogen peroxide to water
Catalase	Enzyme	Cytosol	Fe	Conversion of hydrogen peroxide to water
Alpha tacopherol	Natural chemical	Membrane	Vitamin E	To help in breaking fatty acid peroxidation chain reaction
Carotene	Lycopene pigment	Membrane	Carotene	To prevent initiation of fatty acid peroxidation chain reactions

normal cells into malignant cells, thereby boosting the immune responses. Glutathione peroxidase requires the mineral selenium as a cofactor.

Glutathione directly combats ROS as lipid peroxides and is involved in xenobiotic metabolism as well. Xenobiotics are the exogenous compounds that are not the part of an organism nor a part of its normal nutrition. Whenever the liver is exposed to xenobiotics, it secretes detoxifying enzymes including cytochrome mingled oxidase, which causes redox degradation. Whenever a person is subjected to excessive levels of xenobiotics, the body produces more glutathione to counteract the effects, which is a vital step throughout the bioremediation phase. As a consequence, it's much less readily accessible to behave as just an antioxidant. The synergistic action of glutathione and vitamin C, on the other hand, works in tandem to combat free radicals (Table 1.3).

Another important endogenous antioxidant is lipoic acid. In the Kreb's cycle, it is mentioned as "thiol" as well as "biothiol," a sulfur-carrying compound that catalyzes the oxidative degradation of α-keto acids, such as pyruvate and α-ketoglutarate. Since these lipoic acid and significantly reduced dihydrolipoic acid can quench free radicals both in lipid and aqueous media, they are very important antioxidants, so they are also known as "universal antioxidants." Lipoic acid can often reveal their antioxidant properties as well by complexing thru pro-oxidant metals. Lipoic acid is generally used to guard against both vitamin E and vitamin C insufficiency. List of some endogenous and exogenous dietary antioxidants along with their cofactors are tabulated in Table.1.4.

Some characteristics of ideal antioxidants [11] are displayed in the following Table.1.5.

Table 1.4 List of free radical scavenging antioxidants

Endogenous Autogenerated Antioxidants	Exogenous Dietary Antioxidants	Proteins with their cofactors
Glutathione, lipoic acid, and N-acetyl cysteine are examples of thiols NADPH and NADH are two oxidative phosphorylation enzymes • Uric acid • Ubiquinone (coenzyme Q10) • Enzymes — Superoxide dismutase (SOD) (cofactors: Copper/zinc and manganese — Catalase dependent (cofactor: Iron) Glutathione peroxidase (cofactor: Selenium)	Polyphenols, such as flavonoids, flavones, flavonols, and proanthocyanidins Beta-carotene and also other carotenoids and oxy-carotenoids, like lycopene and lutein Vitamin C (ascorbic acid) Vitamin E (tocopherol)	*Cu-based proteins* • Metallothionein • Albumin • Ceruloplasmin *Fe-based proteins* • Myoglobin • Ferritin • Transferrin

Table. 1.5 Characteristics of Ideal Antioxidants

Ideal antioxidants • Neutralizes a free radical • Prevents the oxidation of a substance, even present in traces • Active ingredients that can slow or prolong the onset of the lipid oxidation process of food products • Not impart an unpleasant taste, odor, or color to the fatty contents • No negative physiological consequences • Efficient even in lower concentrations • Soluble in fat (some in aqueous media as vitamin C) • Easy to accomplish • No degradation during production • Readily available	

3.3.4 Classification on the Basis of the Size of the Antioxidant Molecules

The antioxidants that are found in our body may also be classified on the basis of the size of the antioxidant molecules as small or large ones:

Antioxidants made of smaller molecules: These are the molecules that serve as antioxidants having lower molecular weights. These may be hydrophilic such as vitamin C, glutathione, minerals, etc. or lipophilic such as vitamin E, carotenes, retinol (Vitamin A), ubiquinol, and lipoic acid.

Antioxidants made of larger molecules: These are higher molecular weight antioxidants such as superoxide dismutase and catalase.

Scheme 1.2 Structure of ascorbic acid

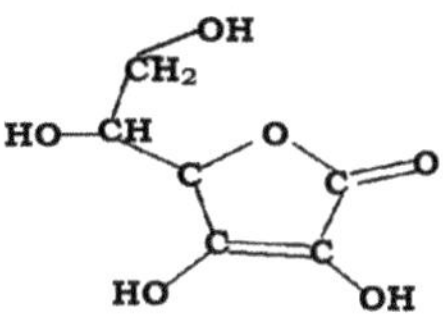

The following are some of the most significant small-molecule antioxidants:

Vitamin C Vitamin C or ascorbic acid is body's one of the most significant hydrophilic water-soluble antioxidant. They usually operate with biological fluids. They efficiently inhibit the emergence of free radicals induced by factors like pollution and tobacco smoke. It has anticancer, anticardiac ailments, anti-arthritis, and antiaging properties. It also aids in the restoration of vitamin E to its active state (Scheme 1.2).

Vitamin C is very beneficial in preventing oxidative damages to tissues. It inhibits the generation of carcinogens. The beneficial and protective effects of fruit and vegetable intake on cancer incidence have been identified in numerous studies. This is largely due to vitamin C's anticancer properties. It has the potential to reduce overall cholesterol levels in the blood, lowering the risk of heart diseases. Anyone with circulating vitamin C levels that become close or below the deficiency limit have a greater risk of cardiovascular disease. Cataract patients were found to have low vitamin C and E concentrations, as well as low plasma vitamin C amounts, in many trials. Large consumption of ascorbic acid rich fruits and vegetables seems to be protective as well [12].

Vitamin E Nuts, seeds, vegetable and fish oils, whole grains (particularly wheat germ), loaded dairy products, and apricots all contain vitamin E, which is fat-soluble tocopherol. It is most efficient chain-breaking antioxidant found in all cell membranes and is primarily contained in adipose tissues, liver, and muscles. It is the body's most important antioxidant, protecting polyunsaturated fatty acids, phospholipids in cellular membranes from oxidation, i.e., important protector towards oxidation or leading defensive player against lipid peroxidation, which results in the formation of reactive unstable intermediate molecules with even more oxygen than normal ones. Vitamin E, in addition to just being a free radical scavenger, boosts the immune system when consumed in large amounts. It even prevents nitrites from being converted to cancer-promoting nitrosamines in the stomach. Its consumption has been linked to a reduced risk of angina and fatal cardiovascular diseases [13].

Carotenoids A class of pigments occurring in plant foods especially fruits and vegetables, that are red, orange, and yellow in color are known as carotenoids. There are only a few carotenoids, such as beta-carotene, which also serve as vitamin A precursors, while others don't do so (Table 1.6).

Preventive Antioxidants Preventive antioxidants are those which are basically enzymes that minimize the formation of initiating radicals. Most common ones are

Table 1.6 List of some antioxidants in edibles

Antioxidants	Edibles
Vitamin A	Milk, egg
Vitamin C	Citrus fruits, broccoli, greens, spinach, bell peppers, berries, grapefruit, pink, mango, papaya, sweet potatoes, tomatoes (canned), watermelon
Vitamin E	Beans, nuts, nut butters, oils, seeds, wheat germ, whole grains
Anthocyanins	Beets, berries, eggplant, grapes, red wine, prunes
Beta-carotene	Apricots, cucumber, spinach, sweet potatoes, root vegetables, acorn squash, green peppers, broccoli, cabbage, plantain, mango, papaya, peaches
Catechins	Apples, beans, tea (black or green), berries
Resveratrol	Blue-berries, peanuts, grapes, red wine
Selenium	Brown rice, chicken, garlic, onions, oatmeal, salmon, tuna, seafood, wheat germ, whole grains
Lutein lycopene	Corn, egg (in yolk), merry gold Grapefruit, pink, tomatoes, watermelon

superoxide dismutase, catalase, glutathione peroxidase, singlet oxygen quenchers, and transition metal chelators (e.g., EDTA).

3.3.5 Classification on the Basis of the Occurrence of Antioxidants

The antioxidants may also be classified, whether they are naturally occurring or of synthetic category:

1. Natural antioxidants: Occurs naturally in biological systems, e.g., nordihydroguaretic acid (NDGA), sesamol, gossypol, and other tocopherols.
2. Synthetic antioxidants: Those which are synthesized in laboratory, e.g., butylated hydroxy anisole (BHA), butylated hydroxy toluene (BHT), propyl gallate (PG), tertiary butyl hydroquinone (TBHQ).

It's important to note that antiradical and antioxidant activity are not quite the same things. Antiradical activity refers to a compound's potential to interact with a free radical, while antioxidant activity refers to the ability to prevent oxidation. As a result, all assays that use a stable free radical, e.g., DPPH, ABTS, provide data on scavenging free radicals or antioxidant capacity, but this behavior does not always equate to antioxidant activity. For the sake of getting the correct statistics on true antioxidant activity in relation to lipids and other factors, it is recommended to carry out the research on natural food stuffs.

References

1. Rice-Evans CA, Miller NJ, Paganga G. Structure-antioxidant activity relationships of flavonoids and phenolic acids. Free Radic Biol Med. 1996;20:933–56.

2. Lü J-M, Lin PH, Yao Q, Chen C. Chemical and molecular mechanisms of antioxidants: experimental approaches and model systems. J Cell Mol Med. 2010;14(4):840–60.
3. Huang, et al. J Agric Food Chem. 2005;53:1841–56.
4. Lü J-M, Yao Q, Chen C. Ginseng compounds: an update on their molecular mechanisms and medical applications. Curr Vasc Pharmacol. 2009;7:293–302.
5. Panchatcharam M, Miriyala S, Gayathri VS, et al. Curcumin improves wound healing by modulating collagen and decreasing reactive oxygen species. Mol Cell Biochem. 2006;290:87–96.
6. Shih PH, Yeh CT, Yen GC. Anthocyanins induce the activation of phase II enzymes through the antioxidant response element pathway against oxidative stress-induced apoptosis. J Agric Food Chem. 2007;55:9427–35.
7. Hurrel R. Influence of vegetable protein source on trace elements and mineral bioavailability. J Nutr. 2003;133:29735–75.
8. Carocho M, Ferreira ICFR. A review on anti- oxidants, Prooxidants and related controversy: natural and synthetic compounds. Screening and analysis methodologies and future perspectives. Food Chem Toxicol. 2013;51:15–25.
9. Rahman K. Studies on free radicals, antioxidants, and co-factors. Clin Interv Aging. 2007;2 (2):219–36.
10. Ratnam DV, Ankola DD, Bhardwaj V, Sahana DK, Kumar MNVR. Role of antioxidants in Prophy- laxis and therapy: a pharmaceutical perspective. Jour- nal of Controlled Release. 2006;113(3):189–207.
11. Shebis Y, Iluz D, Kinel-Tahan Y, Dubinsky Z, Yehoshua Y. Natural antioxidants: function and sources. Food Nutr Sci. 2013;4:643–9.
12. Buettner GR, Jurkiewicz BA. The ascorbate free radical as a marker of oxidative stress: an EPR study. Free Radic Biol Med. 1993;14:49–55.
13. Buettner GR. The pecking order of free radicals and antioxidants: lipid peroxidation, α-tocopherol, and ascorbate. Arch Biochem Biophys. 1993;300:535–43.

Chapter 2
Chemistry of Natural Super-antioxidants

Abstract **Phytochemicals** are the natural super-antioxidants synthesized by plants through various metabolic reactions and thus are plant products. Depending on the nature of metabolic reactions, they may be categorized as primary or secondary. Studies have shown that the diets that include fruits and vegetables well protect our body from various chronic diseases. Due to this fact, fruits and vegetables are recommended to be a part of dietary intake. These pulpy nutrients include polyphenols having most plenteous antioxidants. Polyphenols are chemically, a varied group of naturally occurring weakly acidic, having multiple phenolic functional groups possessing aromatic rings bearing hydroxyl substituents. Accordingly they are classified as flavonoids, phenolic acids and phenolic alcohols and less common lignans and stilbenes. The phytochemicals, in addition to having beneficial properties, are also responsible for the colors of fruits and vegetables. The extraordinary antioxidant activity possessed by polyphenols is confirmed through various researches. Flavonoids act as natural super-antioxidants, especially by the rummaging various oxidizing species, superoxide anion, hydroxyl radical, and peroxyl free radicals. They perform their duties in multifunctional mode of activity. They may serve as reducing agents by donating hydrogen atoms or may act as quenchers for singlet oxygen. It means that the antioxidant activities performed by the phenolic compounds of plants are mainly due to their capabilities of encompassing in redox reactions.

Keywords Phytochemicals · Polyphenols · Flavonoids · Antioxidant activity

1 Introduction

1.1 Phytochemicals

In paving way to nano-antioxidants, it is necessary to keep eye on natural antioxidants rather super-antioxidants. We are blessed for that, for our survival and stay healthy, nature has provided abundance of oxidation antidotes, consequently the name, antioxidants are given to them. We have introduced antioxidants in earlier

S. Sundaram Sanjay, A. K. Shukla, *Potential Therapeutic Applications of Nano-antioxidants*, https://doi.org/10.1007/978-981-16-1143-8_2

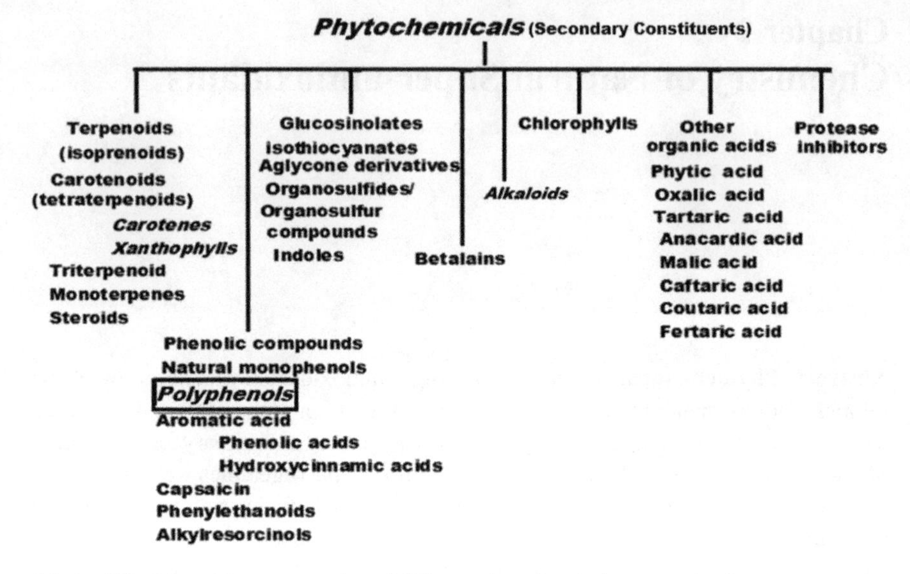

Scheme 2.1 Schematic representation of different phytochemicals present in plants

chapter. There are number of proteins, carbohydrates, and fats within our body which play their important role in its building. As said earlier, these need to be protected from oxidation. Fruits and vegetables combat oxidation directly. These naturally occurring edibles contain abundance of vitamins, minerals, and phytochemicals having natural antioxidant powers. Therefore it will not be an exaggeration to call them super-antioxidants.

According to worldwide dietary recommendations, one should have ensure regularly at least five to ten servings of disease-fighting antioxidant fruits and vegetables intake. These fruits and vegetables contain some essential dietary contents which are physiologically and biologically very important, considered as bioactive substances. Since chemicals from *phytos,* so named as "**phytochemicals**" (Scheme 2.1).

Depending on the nature of metabolites and metabolic reactions, they may be categorized as primary or secondary. Within primary category, there comes amino acids, proteins, some sugars, purines and pyrimidines of nucleic acids, chlorophyll, etc. The other phytochemical products such as alkaloids, terpenes, and phenolics come under secondary constituent category [1].

1.2 Polyphenols

Constituents wise polyphenols are the largest class of phyto-products recognized in the plant kingdom. Their occurrence is not limited to any specific category. They found in significant amounts even in commonly consumed fruits and vegetables.

Many specific studies have shown that their dietary consumption is must for the protection of our body against chronic diseases [2]. This protection, of course, to some extent, is because of the consumption of valuable nutrients containing polyphenols as they have plenty of antioxidants. Since from last few years, considerable attention and awareness was laid down among agro and food scientists, nutritionists, and related industrialists and consumers on the molecules having antioxidant activities such as polyphenols, etc.

Antioxidants as mentioned earlier prevent from innumerable diseases related to oxidative stress, such as inflammation, cancers, cardiovascular diseases, and many others [3, 4]. Through number of experimental data, it is now confirmed that the polyphenol molecules exhibit several biological activities that include radical scavenging and antioxidant activities that take account of anti-inflammatory, antimutagenic, anticancer, anti-HIV, anti-allergic, antiplatelet activities [5].

Thousands of molecules with a polyphenolic assembly (aromatic rings having many hydroxyl groups on) have been recognized and documented in edible plants. Being secondary metabolites of plants, these molecules are usually involved and secreted as result of defense mechanism combatting against ultraviolet radiation or any other pathogenic or stressful conditions. These defensive outcomes constitute almost 9000 currently known compounds and the number of this complex family members is still mounting, with varied structures, sizes (ranging from monomers to polymers), and properties. Biological properties such as bioavailability, antioxidant activity, an explicit interaction with cell entities and enzymes largely depend on the structural chemistry of polyphenols. Generally, organoleptic properties of edibles such as color, odor, taste, and bitterness depend on the basic phenolic structure. Although, physical factors such as pH, temperature, and light also influence their properties. Moreover, the quality of fruits and vegetables are largely decided by the family of polyphenols they belong, along with their neutral-ceutical properties. The concentration and structural criteria of bioactive compounds present in edibles decide the efficiency and effectiveness of its antioxidant action [6]. This in turn decides the quantity and composition of phenolic compounds in fruit-based edibles, which depends on various aspects of growing conditions as well as on technological processes that include production area, yield of fruit, kind and species, varieties, geographical heritage, and maturity time.

1.3 Classification and Chemical Structure of Polyphenols

Classification of polyphenols could be done according to the number of phenolic rings they possess and the functional elements through which rings used to get bound together. The important categories of polyphenols are mostly flavonoids, phenolic alcohols, and phenolic acids while the less common are lignans and stilbenes (Scheme 2.2).

The flavonoids have been identified as the biggest category of polyphenols occurring naturally containing almost 9000 dissimilar flavonoids and with their

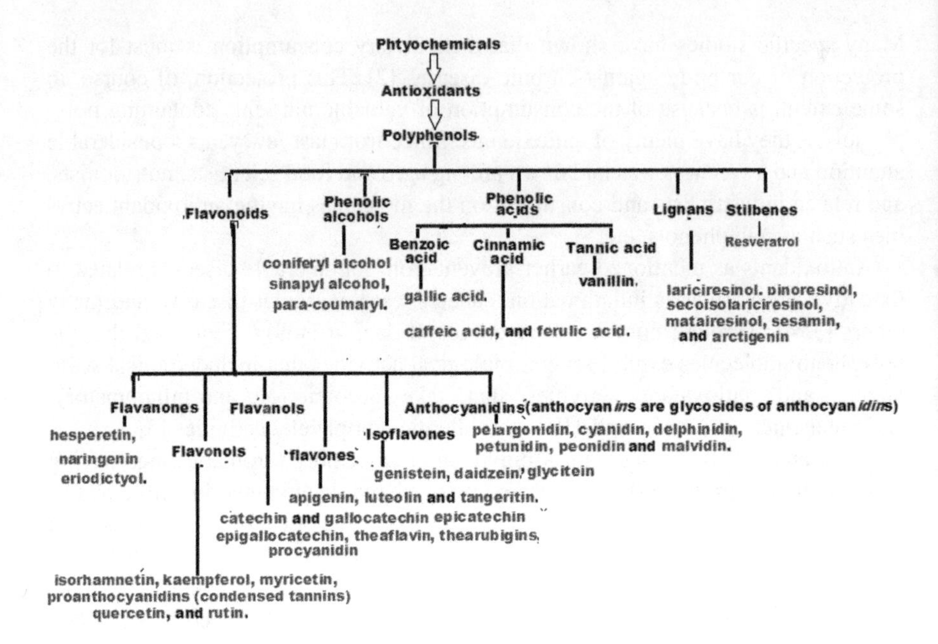

Scheme 2.2 Classification of different polyphenols present in plants

Scheme 2.3 Basic structure of flavonoid

enormous structural diversities associated in these compounds, their list is constantly growing [7]. All flavonoids comprise of two benzene rings connected through a pyran ring having an oxygen atom with a flavan nucleus as a basic unit . The flavonoids can be differentiated by the oxidation level of the basic 4-oxoflavonoid nucleus (the z ring) [8]. They all have a common chemical structure containing two aromatic rings (X and Y) that are connected via a 3-carbon atom forming an oxygenated heterocycle (ring Z). These contain several phenolic hydroxy functional groups attached to ring structures, designated as X, Y, and Z (Scheme 2.3).

In plants, flavonoids generally occur as glycosides. Flavonoid becomes less reactive due to the glycosylation effect, and becomes more water soluble, so that it can be easily stored in the cell sap and vacuoles. Most of the structural diversities in flavonoid category rise due to the various types of substitution reactions in its rings, viz., hydroxylation, methoxylation, glycosylation, etc.

They can be further sub-divided according to the type of heterocyclic ring they possess, as flavonols, flavanols (catechins), flavones, flavanones, isoflavones, and anthocyanidins (Fig. 2.1). Along with the antioxidant potentiality, flavonoids may display certain more interesting biogenic activities, viz., antimutagenic activity,

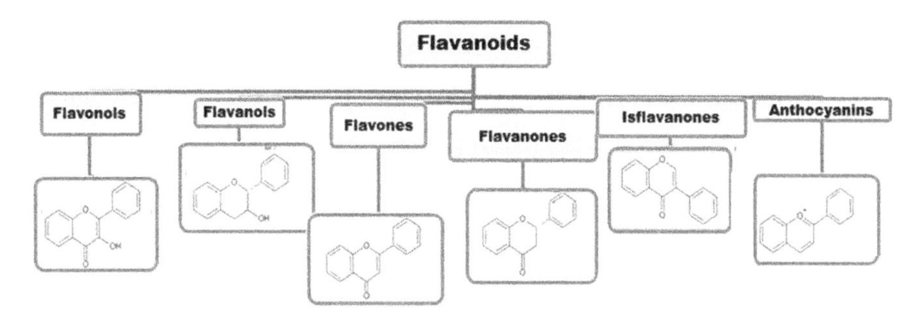

Fig. 2.1 Flavonoids' sub-classification

antimicrobial properties, tumbling the risk of cardiovascular diseases, anti-proliferative action on tumor cells, atherosclerosis protection, hair energizer, serving as hormonal source for women during her natural menopause [9].

The most abundantly found flavonoids present in our diets are the flavonols. Quercetin and kaempferol are its important representative members. The other sub-member, flavones are relatively less commonly found in fruit and vegetables than flavonols. The glycosides of luteolin and apigenin are its major constituents. Till date more than 5.000 flavonoids are known. Some of the important ones are mentioned below:

- Flavonoids found in citrus family are generally rutin, naringin, limonene, quercetin, and hesperidin. In yellowish-green onions, broccoli, cherries, grapes, apples, and red cabbage, the flavonoid present is quercetin. These flavonoids perform their maximum antioxidant activity through various multistep mechanisms. Quercetin decreases interestingly the skin damages caused by the exposure of sun's UV radiations. But its skin-protective application is restricted because of its poor subcutaneous penetration limits [10]. Hesperidin is another wonder compound found in the peels of oranges and lemons. Bitter taste of number of fruits is due to narangin. As name suggests, limonene has been impounded from lemon and lime.
- Genistein and daidzein are the isoflavone flavonoids, isolated from soy foods such as soy milk, tofu, tempeh, bean textured vegetables, and protein.
- Anthocyanins are basically pigments found dissolved in vacuolar cell sap of the epidermal tissues of flowers and fruits, to which they impart a pink, red, blue, or purple color [11]. They exist in different chemical forms, both colored and uncolored, according to the pH of the medium. Although they are highly unstable in the aglycone form anthocyanidins, but when they are in plants, they are resistant to light, pH, and oxidation conditions that are likely to degrade them. Degradation is prevented by glycosylation, generally with a glucose at position 3, and esterification with various organic acids (rubbing citric and malic acids on the exposed surface) and phenolic acids. In addition, anthocyanins are stabilized by the formation of complexes with other flavonoids (co-pigmentation). Anthocyanidins are dark plant pigments responsible for the red and bluish red

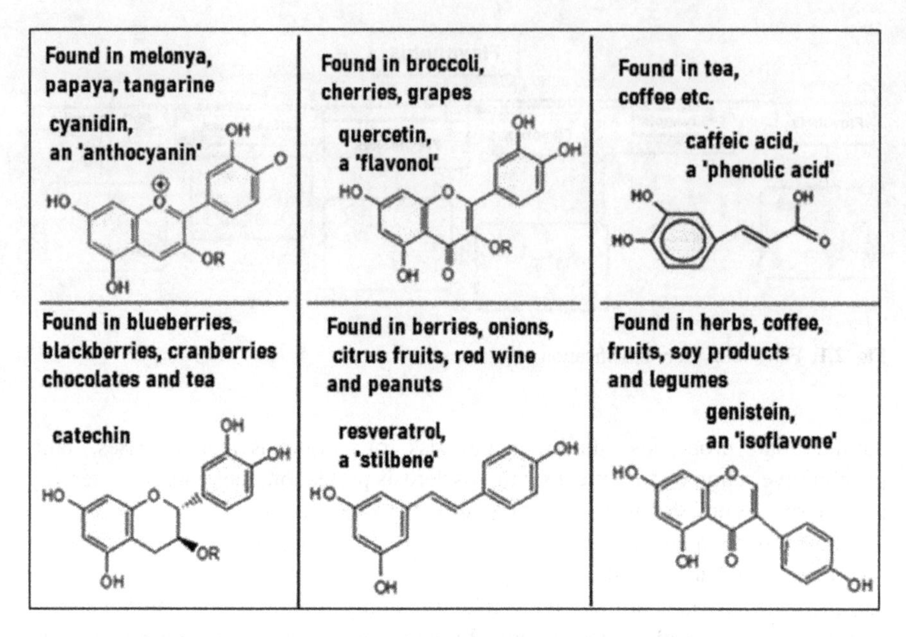

Fig. 2.2 Structures of some polyphenols along with their sources

color of cherries while proanthocyanidins are primarily found in grape seeds, red wine, and pine bark sea extract.

- Lignans are formed of 2-phenylpropane units. Stilbenes are comparatively less common. Resveratrol is a most striking example of this category (Fig. 2.2). It is considered to be an important antioxidant from the family of stilbene and is abundantly found in the skin of grapes, nuts, and berries. Resveratrol shows geometric isomerism, but only trans-resveratrol presents several biological effects, such as anticancer, antiaging, and antioxidant activities. It is an interesting drug to be incorporated in dermal products. Resveratrol has poor oral bioavailability, short half-life, and is extensively metabolized in the body [12].
- The rhizomes of *Curcuma longa* provide a very important polyphenol, curcumin. It is most commonly used in traditional medications and cosmetics for its aesthetic yellow color. Studies have shown that curcumin is very poorly absorbed in the gastrointestinal tract. But its potency increases thousand times synergistically with a ppm presence of black pepper. Curcumin is a potent free radical scavenger quenching superoxide anions, singlet oxygen, and hydroxyl radicals and inhibiting lipid peroxidation. A good candidate for a topical antioxidant should fulfil two conditions: (1) the candidate should permeate through the *stratum corneum* and (2) reach the deeper cutaneous layers without significant leakage into systemic circulation [13, 14].
- The flavonoid that gives value addition to fruits such as grapes, raspberries, and vegetables is "Ellagic acid" [15].
- Green tea and black tea are good sources of "Catechins" [16].

- Kaempferol is profusely found in broccoli, radish, red beets class of vegetables [17].

During the course of interaction with numerous molecules, these natural molecules interact and modify their properties for giving better positive results. Obviously, the color of new herbal product will depend basically on its conjugative bonds initially present in the chemical. During the processing, storage, and aging, there occurs some series of color changes due to the resulting condensation reactions exhibited by the phenolic compounds. Polyphenols are therefore the main culprit for the changes usually observed during storage and aging of fruits and vegetables. Structures of few more polyphenols along with their sources are given in the following Fig. 2.2.

2 Color of Fruits and Vegetables

The phytochemicals not only have beneficial properties, but are also responsible for the colorations of fruits and vegetables [18, 19] (Cámara et al.). The consumption of brightly colored fruits and vegetables is highly recommended for healthy diet due to the presence of specific phytochemical compounds (Fig. 2.3). There are number of phytochemicals that work together synergistically to give different color combinations along with protective measurements of our health. We generally classify colors of fruits and vegetables into seven rainbow groups, namely, green, orange, red, red-purple, orange-yellow, yellow-green, and white-brown, according to the

Fig. 2.3 Antioxidant bearing colorful naturals

Table 2.1 Color classification pattern of fruits and vegetables with the responsible phytochemicals

Rainbow Colors	Phytochemicals (pigments) accountable for the color	Phytochemicals accountable for the properties	Few examples
Green	Glucosinolates, Saponins, β-carotene. Lutein, zeaxanthin.	Chlorophyll as well as folacin, folates, calcium, iron, magnesium, nitrates, glutathione, saponins, ascorbic acid—Vitamin C, caffeic acid, *para*coumaric acid, ferulic acid, chlorogenic acid, sulforaphane, tannins, (sulfur-containing derivatives of amino acid.)	Broccoli, kale, cauliflower, Brussels sprouts and cabbage, green peas, leafy spinach, kiwifruit
Orange	Carotenoids (carotenoids α- y β-carotene) β-cryptoxanthin, Curcuminoids	The pigments mostly belong to the isoprenoid lipids category and derive their color from conjugated C-C alternative double bonds in their structures.	Carrot, mango, pumpkin, oranges, papaya, peaches
Red	Lycopene,β-carotene, lutein, Fisetin, flavones, Phloretin, quercetin, kaempferol,Ellagitannins	Mostly the carotenoids, such as α- and β-carotene	Tomato, strawberries, cherries, radishes. Apple
Reddish-purple	Anthocyanins, Anthocyanidins, Proanthocyanidins	Flavonoids	Berries (black, blue), raspberries, plums, black current, grapes
Yellowish-Orange	Anthocyanins, β-Cryptoxanthin,	Flavonoids, carotenoids, α-, β-, and γ- carotene.	Melons, peaches, papaya, orange, tangerine, honeydew
Greenish-yellow	Zeaxanthin, lutein, quercetin, bioflavonoids, bromelain, gingerol,	Xanthophylls (carotenoid)	Spinach, avocado, melon, apples, broccoli, cherries, grapes, lemons, pine apple, papaya, sweet corn,
Brown/ white	Allicin, indoles, and isothiocyanates	3,3'-Diindolylmethane syn-propanethial-S-oxide	Cauliflower, garlic, ginger, mushrooms, onions, potatoes, bananas, brown pears, white peaches, nectarines and dates

phytochemicals they contain. The color classification of fruits and vegetables along with their responsible phytochemicals is briefly summarized in the following Tables 2.1.

3 Antioxidant Activity of Polyphenols

The medicinal properties of plants are generally exhibited due to the presence of polyphenols constituents present as active substances, which modulate the activity of concerned enzymes and cell receptors. Flavonoids inhibit lipid peroxidation and lipoxygenases acting as an antioxidant. Number of studies have been confirmed that the flavonoids perform their antioxidant activity specifically through redox reactions, involving sifting of various oxidizing species including superoxide anions, hydroxyl radicals, and peroxyl free radicals, following different mechanisms, in which they act sometimes as reductants, hydrogen donor antioxidants, or quenching singlet oxygen, thus contributing overall antioxidant activities of plants primarily by exhibiting their redox characteristics, and neutralizing lipid free radicals and checking the breakdown of hydrogen peroxides into free radicals [20].

As discussed earlier, free radicals may be exogenous or endogenous in their origin. Endogenously they are produced within the body through several metabolic pathways, may be as a result of mitochondrial respiration or any enzymatically catalyzed oxidation reactions. These random free radicals do not sit idly, instead, freely react with nearest neighboring compounds by capturing their electrons to make itself stable and making the reactant unstable, initiating a radical chain reaction, e.g., lipid peroxidation, which results in the impairment that may even lead to the rupture of cell membranes. Since free radical reactions are usually very fast, their kinetics can only be studied via reactants' diffusion rates. Similar oxidative mechanism may also happen to proteins and DNA, causing in cellular dilapidation [21]. The mechanism of free radical production was discussed in previous chapter. To reverting back the action of free radicals, the enzymatic defense warriors, i.e., antioxidants, come into action, with their oxidoreductases weapons, as superoxide dismutases, peroxidases, catalase, and glutathione peroxidase along with non-enzymatic antioxidants, namely reduced form of glutathione, ascorbic acid, α-tocopherol, β-carotene, polyphenols, etc. [22] work endogenously. The additional high antioxidants intake via fruits and vegetables to our daily regular diet provides high nutritional value to make our defense systems highly active. The high antioxidant potentiality of polyphenolic compounds is primarily related to their free radicals scavenging ability that is produced as a result of oxidative stress.

All the edible resources whether it may be polyphenols or flavonoids, they should have capability to perform antioxidant activity. For that, two basic criteria should be fulfilled by the antioxidant entity. Firstly, it should be capable to avert or hinder free radical-enabled oxidation or autoxidation by delaying the process even at very low concentrations and, secondly, the formation of free radicals occurring via chain reactions should be terminated through any process, that may be the formation of hydrogen bonding or hydrogen donation to check further oxidation of the molecule [23]. As we know that, the configuration and structure of a chemical compound decides its chemical behavior, thus the antioxidative behavior of polyphenols or flavonoids also seeks some basic parameters which may be attributed to the nature of compound, number and position of the hydroxyl groups present, and the degree or amount of methoxylation in their chemical structures. These characteristics gives rise to the ability to serve them as free radical hunters. Availability of this phenolic hydrogens as hydrogen donating groups makes them excellent antioxidants. As per Bors and their associates [24], the radical-gulping capability of flavonoids has three important structural determinants. That are: (a) the dihydroxy catechol structure at y ring; (b) the conjugated double bonds with 4-oxo functional group at z ring; and (c) the presence of hydroxyl groups at second, third, or fifth positions at rings x, y, and z (Scheme 2.3). It is thus obvious that this basic structural criteria is necessary for the electron bestowing capability whether it is flavonoids to peroxidases, i.e., to display antioxidant activity or radical scavenging efficacies. Out of the number of hydroxyl groups present in the structure, the 3-OH groups play the utmost substantial determining factor for electron-serving hub [25]. To perform antioxidant activity by scavenging superoxide radical, structural modification of edible polyphenols may occur within the digestive track specially in intestine and during the metabolic activities procured by liver, which places a considerable effect on the consequent in vivo biological activity. Moreover, the structural modifications derived from dimerization can also increase the superoxide scavenging capacity but the ultimate effect would largely be influenced by the type of bonding between the monomer entities [26].

Now it is clear that the total number of hydroxyl groups in the configuration of the flavonoids places a significant impact and plays an important role in the reaction mechanism of antioxidant activity. Where the presence of hydroxyl group on ring y plays an important role in ROS hunting t hough the substituent present at the rings x and z do not put much influence on the rate constants for the superoxide anion radical rummaging activity [27].

For the prediction of free radical scavenging activity of flavonoids can be easily done with the reduction potentials of their antioxidant reactivity. Reduction potentials of flavonoid radicals are much lower on comparing it with the superoxide radicals or alkyl peroxyl radicals. This indicates that the flavonoids have much more capabilities to deactivate the ROS species and thwart the venomous consequences of their interactions. Similar results have been observed in number of in vitro studies. It has been established that the antioxidant potency of polyphenols is due to their scavenging capability of free radical and slowing down the oxidation of low density lipoproteins, otherwise may lead to chronic diseases [28, 29].

References

1. Walton NJ, Brown DE. Chemicals from plants: perspectives on plant secondary products. London: Imperial College Press; 1999.
2. Arts ICW, Hollman PCH. Polyphenols and disease risk in epidemiologic studies. Am J Clinival Nutr. 2005;81:317–25.
3. Steinmetz KA, Potter JD. Vegetables, fruit and cancer prevention: a review. J Am Diet Assoc. 1996;96:1027–39.

4. Rimm EB, Ascherio A, Giovannucci E, Spiegelman D, Stampfer MJ, Willett WC. Vegetable, fruit and cereal fiber intake and risk of coronary heart disease among men. J Am Med Assoc. 1996;275:447–51.
5. Harbone JB, Williams C. Advances in flavonoid research since 1992. Phytochemistry. 2000;55:481–504.
6. Pietta PG. Flavonoids as Antioxidants. J Nat Prod. 2000;63:1035–42.
7. Hertog MLG, Hollman PCH, Katan MB. Content of potentially Anticarcinogenic flavonoids of 28 vegetables and 9 fruits commonly consumed in the Netherlands. J Agric Food Chem. 1992;40:2379–83.
8. Aherne SA, O'Brien NM. Nutrition. 2002;18:75.
9. Yao LH, Jiang YM, Shi J, Tomás-Barberán FA, Datta N, Singanusong R. Flavonoids in food and their health benefits. Plant Foods Hum Nutr. 2004;59:113–22.
10. Kitagawa S, Tanaka Y, Tanaka M, Endo K, Yoshii A. Enhanced skin delivery of quercetin by microemulsion. J Pharm Pharmacol. 2009;61:855–60.
11. Mazza G, Maniati E. Anthocyanins in fruits, vegetables, and grains. Boca Raton: CRC Press; 1993.
12. Hung CF, Lin YK, Huang ZR, Fang JY. Delivery of resveratrol, a red wine polyphenol polyphenol, from solutions and hydrogels via the skin. Biol Pharm Bull. 2008;31:955–62.
13. Fang JY, Hung CF, Chiu HC, Wang JJ, Chan TF. Efficacy and irritancy of enhancers on the *in vitro* and *in vivo* percutaneous absorption of curcumin. J Pharm Pharmacol. 2003;55:593–601.
14. Suwannateep N, Wanichwecharungruang S, Haag SF, Devahastin S, Groth N, Fluhr JW, Lademann J, Meinke MC. Encapsulated curcumin results in prolonged curcumin activity *in vitro* and radical scavenging activity *ex vivo* on skin after UVB-irradiation. Eur J Pharm Biopharm. 2012;82:485–90.
15. Aiyer HS, Srinivasan C, Gupta RC. Dietary berries and ellagic acid diminish estrogen-mediated mammary tumorigenesis in ACI rats. Nutr Cancer. 2008;60:227–34.
16. Musial C, Kuban-Jankowska A, Gorska-Ponikowska M. Beneficial properties of green tea Catechins. Int J Mol Sci. 2020;21(5):1744. Published 2020 Mar 4. https://doi.org/10.3390/ijms21051744.
17. Park S, Arasu MV, et al. Metabolite profiling of phenolics, anthocyanins and flavonols in cabbage (*Brassica oleracea* var. *capitata*). Indus Crops Products. 2014;60:8–14.
18. Cámara Hurtado M, de Cortes Sánchez Mata M, TorijaIsasa M. Frutas y verduras, fuente de salud. Instituto de SaludPública. Consejería de Sanidad y Consumo; 2003.
19. Heber D, Bowerman S. Applying science to changing dietary patterns. In: American Institute for Cancer Research 11th annual research conference on diet, nutrition and Cancer. Bethesda: Am Soc Nutr Sci; 2001.
20. Javanmardi J, Stushnoff C, Locke E, Vivanco JM. Antioxidant activity and total phenolic content of Iranian Ocimum accessions. Food Chem. 2003;83:547–50.
21. Halliwell B, Aeschbach R, Löliger J, Aruoma OI. The characterization on antioxidants. Food Chem Toxicol. 1995;33:601–17.
22. Mallick N, Mohn FH. Reactive oxygen species: response of algal cells. J Plant Physiol. 2000;157:183–93.
23. Shahidi F, Wanasundara PKJ. Phenolic antioxidants. Crit Rev Food Sci Nutr. 1992;32:67–103.
24. Bors W, Heller W, Michel C, Saran M. Flavonoids as antioxidants: determination of radical-scavenging efficiencies. Methods Enzymol. 1990;186:343–55.
25. Takahama U, Egashira T. Peroxidase in vacuoles of Viciafaba leaves. Phytochemistry. 1991;30:73–7.

26. Tsajimoto Y, Hashizume H, Yamazaki M. Superoxide radical scavenging activity of phenolic compounds. Int J Biochem. 1993;25:491–4.
27. Amić D, Davidović-Amić D, Bešlo D, Trinajstić N. Structure-radical scavenging activity relationships of flavonoids. Croatian Chem Acta. 2003;71(1):55–61.
28. Jovanovic SV, Jankovic I, Josimovic L. Electron-transfer reactions of alkyl peroxy radicals. J Am Chem Soc. 1992;114:9018–22.
29. Wardman P. Reduction potentials of one-electron couples involving free radicals in aqueous solution. J Phys Chem Ref Data Ser. 1989;18:1637–755.

Chapter 3
Nano-antioxidants

Abstract All the living beings suffer directly or indirectly through the harmful effects of the free radical reactive oxygen species causing serious damages to the cells and organelles by attacking constituents of cell membrane, the unsaturated fatty acids. Presence of antioxidants defend against free radical damage. The antioxidants are manmade or natural, play a very important role against the free radicals, pathogens, and diseases by providing a physiological shield against oxidative stress. Because of the carcinogenicity of the synthetic antioxidants and increased reactivity in nano-size, now the scientists have progressed in evaluating the antioxidant activity of nanomaterials such as capped nano-gold, silver, ceria, yttria, and many natural polymers including fullerenes. There are number of methods for the evaluation of the antioxidant activity of metals or compounds. Among those some important methods are discussed in this chapter to brush up with antioxidants. The antioxidant mechanism and its cytotoxicity are also discussed briefly.

Keywords Antioxidants · Free radical · Metal nanoparticles · Lipid peroxidation

1 Introduction

Traditional antioxidant therapies have not been yet considered to be a sure shot option for the line of treatment for various diseases resulted due to oxidative stress, as the traditional medicines become incapable of crossing the most protected blood–brain barrier. To overcome such constraint, nanoparticle antioxidants emerged as an alternative expansion of antioxidant treatments to inhibit and handle the diseases related to oxidative stress with proper precaution and care. Due to their compatible size, nanoparticle antioxidants are believed to have proper and unwavering associations with biomolecules and proven to have extra efficiency and effectiveness to counteract the damages caused by free radicals.

S. Sundaram Sanjay, A. K. Shukla, *Potential Therapeutic Applications of Nano-antioxidants*, https://doi.org/10.1007/978-981-16-1143-8_3

1.1 A Brief Introduction of Nanoparticles

Before discussing nano-antioxidants, it is indispensable to give a brief introduction of nanoparticles. Goggling with nanoparticles is nanotechnology, term refers to the build up materials right from the lowest level meticulously with great precision. It means that fabricating and designing such materials which has at least one of its dimensions at nanometer (10^{-9} m) scale level or less comes within the purview of nanotechnology. Within this purview, a size-dependent property dominates. So the functions are carried out accordingly by manipulating and controlling the atoms and molecules at the nanometer scale levels. There is involvement of all the science trio subjects, i.e., physics chemistry, and biology, together churned for the synthesis of materials and to get desired structures, components, devices, and systems for any purposeful applications.

Since nanomaterials have their unusual size and developed physicochemical properties due to that, these materials offer a huge arena of advantages, which revolutionized the whole world. In the dominion of nano-biotechnology, the scientists including chemists, physicist, biologists, and engineers have cosseted their research work in full swing for the welfares of society and human healthcare, which include fabrication, modelling, designing medicines, sensors, and many more.

There are number of ways through which nanomaterials can be synthesized. It includes wounding, extruding, fragmenting, thrashing, beating, ball milling, co-precipitation and through numerous other means using bottom-up or top-down approaches. Usually different preparatory methods of nanomaterials make their structure and morphology different, which can be modified according to the requirements. The atoms in nanomaterials align themselves in space in different ways, different from their bulk form of the same material. Due to which they develop significant variation in their properties.

Nanomaterials may be zero dimensional (0D) quantum dots having three dimensions confined at the nanometer scale range or one dimensional (1D) with two dimensions at the nanometer scale level, as nanotubes, nanorods, nanowires, etc., or two dimensional (2D) having one dimension at nanometer scale range, or three dimensional (3D), as nano-phase things at colloidal range with different structural and morphological variations. Nanomaterials may be crystalline or amorphous in its appearance [1] (Fig. 3.1).

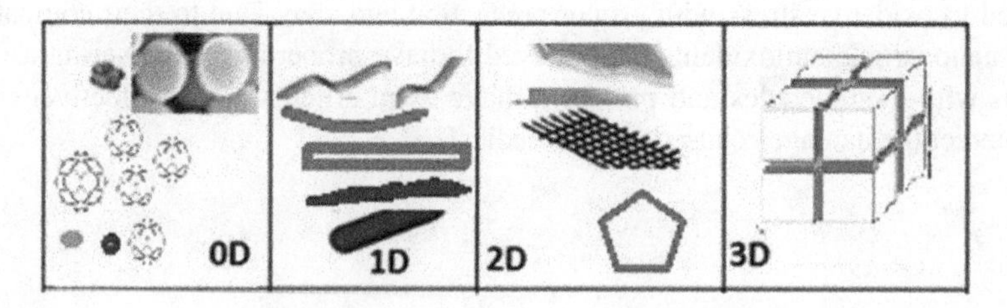

Fig. 3.1 Nanomaterials at different dimensions

1.2 Unique Traits of Nanomaterials

With the reduction in the size of a material proceeding towards nanoscale level, their properties such as melting point, fluorescence, electrical conductivity, magnetic permeability, and chemical reactivity change as a function of the size of the particle and differ dramatically from its bulk form. The rationale behind for this charismatic change in the properties of nanomaterial things may be accredited firstly owing to the discreteness of energy levels. The energy levels in atoms or molecules are quantized. Because of the confinement of the electronic wave function three-dimensionally, the discreteness of the electronic energy levels gives rise to finite density of states. This phenomenon is called as quantum confinement. As a result of quantum confinement, electrical as well as magnetic properties of the material change. Another rationale for the change could be attributed to the tremendous increase in the ratio of surface area to volume at nanoscale (Fig. 3.2). As the particle size reduces, the percentage of atoms at the surface increases in comparison to the atoms present at the center. This increased surface to volume ratio in turn increases the amount of exposed surface area and thus surface energy, which causes the increase in the rate and kinetics of a chemical reaction and changes the surface-dependent properties such as melting point, capillarity, and adhesive properties. The increased surface to volume ratio as a function of particle size is clearly depicted in Fig. 3.2.

1.3 Synthesis of Nanoparticles

Nanotechnology can be done either by "top-down" technology, which means building something by starting with a larger piece and carving away material, which can be compared by the cutting, chipping, pounding, extruding of traditional methods of forming materials. The alternative technique for the formation of nanomaterials from the atomic or molecular structures is known as "bottom-up" technique, which means

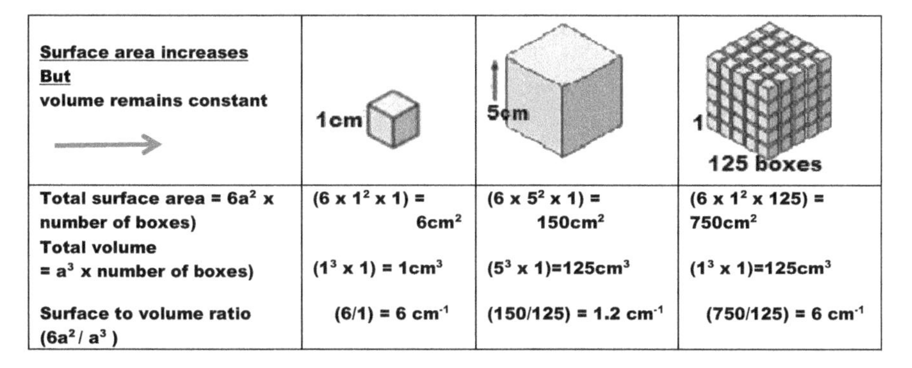

Surface area increases But volume remains constant →	1cm	5cm	1 125 boxes
Total surface area = $6a^2$ x number of boxes) Total volume = a^3 x number of boxes)	($6 \times 1^2 \times 1$) = $6cm^2$ ($1^3 \times 1$) = $1cm^3$	($6 \times 5^2 \times 1$) = $150cm^2$ ($5^3 \times 1$)=$125cm^3$	($6 \times 1^2 \times 125$) = $750cm^2$ ($1^3 \times 1$)=$125cm^3$
Surface to volume ratio ($6a^2/a^3$)	(6/1) = 6 cm^{-1}	(150/125) = 1.2 cm^{-1}	(750/125) = 6 cm^{-1}

Fig. 3.2 Representation of increase in surface area as the decrease in the particle size

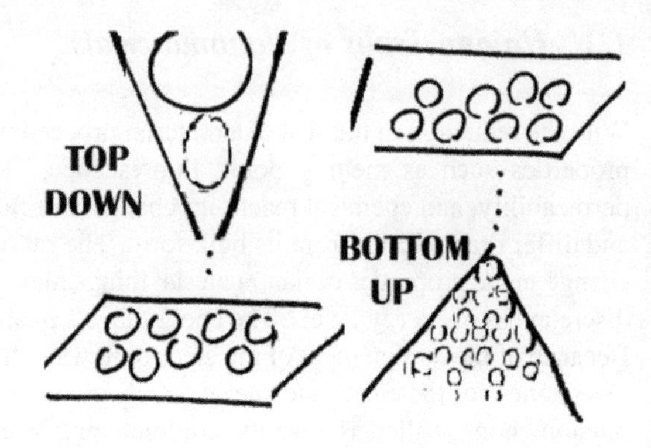

Fig. 3.3 Representation of top-down and bottom-up approaches

building something by putting together smaller pieces, i.e., beginning from the atomic level to nanoscaled structures (Fig. 3.3).

The selection of suitable method for the preparation of nanoparticles depends on the type of application whether it is used for any physical, chemical, or biological applications. Accordingly its physicochemical characters have to be planned. The primary synthetic processes of nanoparticles include mechanical grinding and ball milling which come under the "top-down" technology. In the case of wet chemical synthetic methods for nanomaterials, sol-gel process and co-precipitation are important ones. Under gas phase synthetic techniques, chemical vapor deposition (CVD), chemical vapor condensation (CVC), sputtered plasma processing, microwave plasma processing, and laser ablation are some important ones. For emulsification, solvent evaporation methods, salting out diffusion method, solvent displacement or precipitation method are important. Solvent displacement method involves the precipitation of a preformed polymer from an organic solution and then the diffusion of organic solvent in the aqueous medium is considered. Nanoparticles prepared without any stabilizer generally tend to agglomerate rapidly into micron to millimeter-sized clusters. With the help of capping agents or stabilizers, the size, reactivity, delivery, and transport of nanoparticles can be facilitated. For biological applications, nanoparticles has to be functionalized to make them biocompatible with hydroxyl-, amino-, or thio- groups capping. Engineered nanoparticles may be produced by number of abovementioned synthetic processes, mainly through controlled chemical reactions, while naturally occurring nanoparticles are generated by the erosion and chemical degradation of plants, clay, etc.

1.4 Characterization of Nanoparticles

Characterization of nanoparticles is very important for their identification, application, and further studies. The particle size, their distribution, morphology, surface charges, etc. are some important parameters for their characterization. By dynamic light scattering (DLS), the nanoparticle size can be determined. Morphology and

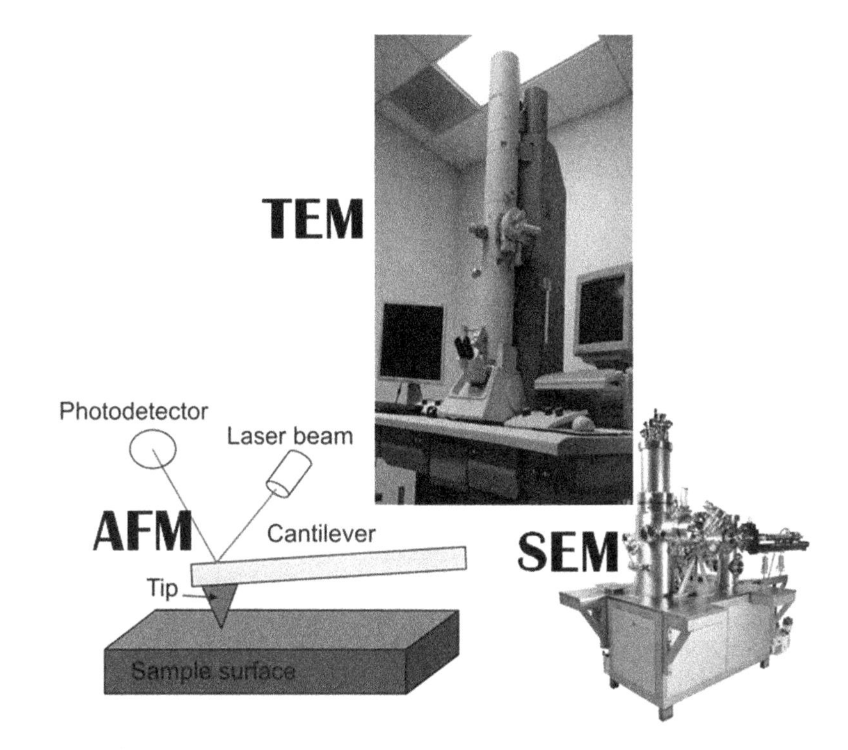

Fig. 3.4 Pictures of TEM, SEM, and AFM

particle size can be determined with the help of advanced microscopic techniques such as scanning electron microscopy (SEM). SEM gives morphological examination by direct visualization. The mean size obtained by SEM can be compared with the results obtained by dynamic light scattering. Transmission electron microscopy (TEM) operates on different principle from SEM, yet it often conveys same type of data. The surface characteristics of the sample can also be obtained with the help of atomic force microscopy (AFM), in which a beam of electrons is transmitted through an ultrathin sample, interacting with the sample as it passes through it. It offers an extremely high resolution to measure the size of the nanoparticle. Basically it scans the samples physically at a very ultramicron miniature level using an atomic scale level probe tip. An AFM creates a structural map of a sample by moving the probe tip on the sample surface and generating figure by the relative forces experienced between sample and the tip (Fig. 3.4).

The nature and intensity of the charge on the surface of nanoparticles in colloidal suspensions plays very significant role. Actually it is the surface charges that determines the potential of electrostatic interaction of nanoparticles with the biological active compounds within the arena of biological environment. The stability of the nanoparticles in its colloidal phase is determined through the measurement of its zeta potential. Though it is not a direct mean for the measurement of the surface charges on the colloids, its measurement directly allows for the predictions about the storage capability and stability of colloidal dispersion. In order to ensure the stability and to avoid agglomeration of the particles, high zeta potential values are preferred, which may be either positive or negative. The extent of surface hydrophobicity can also be predicted with the zeta potential values. In the case of core-shell topologies, with the help of zeta potential values, we can easily get information about the nature of the encapsulated core material. Surface hydrophobicity can be determined with the help of hydrophobic interaction chromatography, biphasic partitioning, adsorption of probes or contact angle measurements. X-ray photon correlation spectroscopy and FTIR spectroscopy help to probe functional groups coated onto the surface of nanoparticles.

2 Nanoparticles as Nano-antioxidants

Today scientists have a clearer picture of how to create nanoscale materials with those properties which are never envisioned before. Recently, researchers are paying considerable attention on the noble coinage metal nanoparticles owing to their extensive applications specially in medicine, biology, optoelectronics, and material science. Applications of nanomaterials in biomedical field specifically rely on magnetic nanoparticles with restricted particle size distribution and manifold enhanced magnetization value. Such magnetic and even non-magnetic nanoparticles have to be coated or surface functionalized to make them biocompatible and no need to say should be nontoxic, and their targeted delivery to their precise ether should be ensured. The stabilization of nanoparticles can be done through various means. The capping tactics could be availed through surface coating with help of specific organic reagents. They may be glycols, polysaccharide, chitosan, or any organic amino acids or inorganic silica, metal oxides or nitrogen, sulfur, oxygen containing compounds [2–6]. These processes functionalize nanomaterials and simultaneously restrict them from aggregation [7, 8]. The greatest advantage of such magnetic and nonmagnetic nanoparticles is that they can hold biomolecules and drugs in this manner and with the help of an external dragging magnetic field can be concentrated to the effected diseased targeted area [9–12]. Recently, progresses have also been attained in the evaluation of antioxidant activity of nanomaterials [13–14].

2.1 Need of Antioxidants

As discussed earlier, our body produces free radicals in a natural way by the metabolic breakdown of our food through chemical reactions involved in digestion. Another uncontrolled source of free radical production within our body is our immune system which generates free radicals as per the need of our body to fight off bacteria or foreign bodies which may cause harmful diseases, i.e., to encounter harmful stimuli, our body's defense system becomes active. During this course, free radicals are generated as by-products. As a result of which too many highly active reactive oxygen and nitrogen species (ROS and RNS) are produced. As long as our body's defense mechanism combats efficiently with these unwanted invasions, our system works normally. The problem arises when the amount of free radicals become higher than the holding capacity of our body. In such cases, free radicals produce serious damages to our body system. Oxygen radicals are electrophilic and form hydroperoxides, which break down into radicals which in turn cause autoxidation of cellular proteins, nucleic acids, and lipids. Hydroxyl, hydroperoxyl, peroxyl, and lipid alkoxyl radicals are mainly responsible for non-enzymatic lipid oxidation, which results in the formation of toxic hydroxy and ketocholesterols, etc. [15] which are damn dangerous to our tissue system. The destructive processes caused by these radicals are called as oxidative stress. In the case of neurodegenerative disorders, generally there occurs death of neuron cell. Oxidative stress evokes toxic reactive oxidative species (ROS) causing to give rise of such phenomenon. Since mitochondrial physiological reactions generate the maximum concentration of reactive oxygen species (ROS), they are also the source of neurodegeneration. These side reactions misguide mitochondria and instigate it to increase the ROS production, which forces to impose oxidative stress on our body producing negative effects allowing pathogens to attack that may result in numerous chronic diseases including inflammation, diabetes, hemorrhagic shock, cataracts, cardiovascular disease, and neurodegenerative disorders leading to body's aging process [16–19].

Through the consumption of antioxidants, we could easily overcome this odd situation. Because antioxidants are the elixir to fight against free radicals. Endogenous antioxidants are produced during the course of normal metabolism in the body, while other lighter exogenous antioxidants can be taken through our regular diet. The specific types we know of are mostly in vitamins and phytochemicals (which come from food sources). There are a lot of different types of antioxidants and each antioxidant has a specific, or preferred, role in helping our body. The role of antioxidants is discussed in previous chapters. The protective role of antioxidants against pathogens has been widely studied [20–22], which has promoted the development of antioxidants for the treatment of diseases associated with oxidative stress.

Till date, a huge number of natural and synthetic antioxidants has been explored to combat and inhibit these oxidation reactions. Moreover, a large number of approaches have been instituted to in vivo *and* in vitro evaluation of the antioxidant activities [23–25].

As discussed earlier, natural antioxidants such as Vitamins (A, C, E) and carotenoids are capable to generate a stable intermediate by accepting an unpaired electron. These intermediates, having considerable half-life time, interact in a controlled fashion, thus preventing autoxidation and the excess energy of electron is dissipated without doing any harm to the tissues. The antioxidative action of phenolic compounds is well documented [26]. Among the various classes of antioxidants, flavonoids, the naturally occurring phenolic compounds brought attention due to their inhibition of lipid peroxidation and lipoxygenase. Polyunsaturated fatty acids through lipid peroxidation generate excessive free radical causing dissemination of oxidative offense system to become active by undergoing production of various types of free radicals via chain propagation. Lastly chain termination occurs through enzymatic interruption by the corrugation of free radical mediated by antioxidants [27]. These mechanisms are discussed in forthcoming chapter in detail.

Due to mitochondrial impairment, numerous problems gets accelerated. They may be aging, neurodegenerative dysfunction, or any other ailment. Under the antioxidant therapeutical efficacy, antioxidants stimulate the brain cells to work against oxidative damages caused by free radicals. Since free antioxidants have to work beyond blood–brain barrier (BBB) in certain critical cases, therefore application of only high doses of molecules have been proven to show curative effect via flowing through the cerebral region. But even then bioavailability of the antioxidant cannot be ensured. Therefore to develop an effective and efficient defensive mechanism for the delivery of antioxidant molecules to the cerebral region has to be chalked out [28]. Moreover, natural antioxidants are very prone to degradation and due to the low absorption and degradation during their delivery their bioavailability becomes limited. Additionally, because of the negative effect on health and carcinogenicity of the synthetic antioxidants such as butylated hydroxyl toluene, tertiary butylated hydroquinone, gallic acid esters, etc. have limited uses. From here, as an elixir, the nanoparticles emerged as the solution to the oxidative stress therapy.

2.2 *Role of Nanoparticles as Antioxidants*

Nanotechnology has now placed its tentacles deep rooted in the field of nanomedicines, whether it is diagnostic, therapeutic, or surgical field. Nano-antioxidants is the curative health concerned branch of engineered nanostructured particles to serve as antioxidants with enhanced characteristics. Sharpe and his co-workers for the first time have coined the term nano-antioxidants. According to them, nano-antioxidants can be categorized into two classes: firstly, the nano-antioxidants possess characteristic inbuilt antioxidant properties having inorganic nanoparticles and secondly, those complexed inorganic nanoparticle antioxidants, which are compelled to perform its function with the help of certain modifications in its nanostructures through nanofunctionalization, viz. nanoencapsulation or nanocapping. There are some natural antioxidants or derived antioxidant enzymes to serve as a carter or antioxidant carriage vehicles to boost the bioavailability of natural antioxidant activity and make provision for the selective distribution for those antioxidants that have low penetration through the cell membranes and cell internalization [29].

Immobilization of natural antioxidants, which are normally organic in nature, can be done more easily than the inorganic nanoparticles, which are mostly thermally stubborn and chemically inert. Furthermore, natural antioxidants coated to nanoparticles provide chemical constancy to the desired antioxidants in biological conditions by delivering and releasing the required antioxidants slowly and unremittingly.

These nanoparticles are generally functionalized with certain bioactive reductants or specific enzymes which serve as antioxidants or function as an antioxidant carriage, thus showing promising performance as therapeutic nanomedicines in oxidative stress eradication, with possible applications in the treatment and prevention of chronic diseases.

On the other hand, most of the antioxidants that have unfavorable physicochemical properties, having extreme conditions regarding to their lipophilicity, hydrophilicity, chemical uncertainty, and low skin penetration, are currently used in a variety of medicinal formulations, limiting their efficacy after topical applications. As a result, nanocarriers including liposomes, niosomes, microemulsions, and nanoparticles have been extensively studied as effective vehicles for drug delivery.

Because of its proclivity for accumulating in enflamed areas of the body, these are recognized as a special commercial carriage system. A greater penetration and efficient liberation of particular therapeutic agents exerting lower risk than conventional therapies is observed when polymeric nanoparticles are targeted to the brain [30]. The capability of nano-encapsulation technology to cross the BBB, as well as its ability to increase drug concentration in cell sap, has been studied successfully. Of course the stability of nanoparticles for drug release and cellular uptake efficiency is essentially determined by their size factor. As a result, nano-antioxidants may serve as effective delivery systems or be used to enhance the treatment's beneficial effects for procurement of the ailments by increasing antioxidant's capability.

In successive sections, various forms of nano-antioxidants along with few examples are discussed, viz. metal/metal oxide nano-antioxidants, non-metal nano-antioxidants, polymeric nano-antioxidants, nanoemulsions, etc (Fig. 3.5).

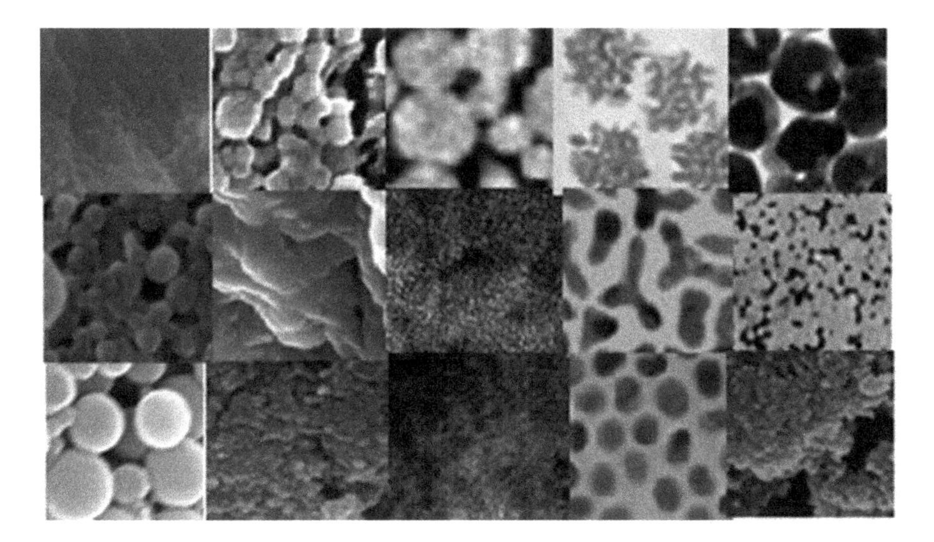

Fig. 3.5 SEM images of some nanomaterials

3 Metal/Metal Oxide Nanoparticles

The noble metal nanoparticles such as copper, silver, and gold have rare unusual optical absorptions, making them important applicatory moieties. Microorganisms, enzymes, and a many more biological processes can be considered for their synthetic purposes.

Some nanoparticles' redox chemistry causes them to become biologically active as antioxidants. Moreover, noble metal nanoparticles, shells like gold and silver may help in protecting the core of the shell. Yin in an editorial comment, emphasized the engineering of nanomaterial with a novel approach to improving antioxidants without changing their central core structure [31]. Radical oxygen species (superoxide radicals, hydroxyl radicals, peroxyl radicals, and so on) as well as non-radical species can cause lipid oxidation, resulting in the formation of hydrogen peroxide, singlet oxygen, ozone, etc. Transition metal ions (Fe^{2+}, Cu^{+}, Co^{2+}) are persisting impurities in food systems and human plasma. The worst thing is that they have potential to accelerate lipid oxidation. Instability and poor water solubility in extreme conditions such as variation in temperature, light, and pH are encountered during the products processing. These variations also have to be tackled with the oxygen radicals present in the gut due to the change in pH, existence of other nutrients, enzymes, or changes during storage period, all factors have to be considered. These factors prevent to exhibit beneficial properties of the components in foods and pharmaceuticals. Nanoparticle-based controlled delivery systems have demonstrated their ability to protect, monitor, and enhance the action of a variety of bioactive compounds [32].

Let us discuss antioxidant properties of some nanoparticles derived from metal, metal oxides, and other than metal-based nanoparticles.

3.1 Gold(Au)

As said earlier, the most widely studied research topic is the protective role of antioxidants against pathogens. It has promoted the development of antioxidants for the treatment of diseases associated with oxidative stress. Gold is a well-known metal that is biocompatible. Historically, colloidal gold was used as a drinkable sol with curative properties for a number of diseases. It has been commonly used as a significant material in the fields of bio-diagnostics, drug/DNA delivery, and cell imaging due to its unusual therapeutic activity, inertness, and nontoxic nature, immune-staining, bio-sensing including pharmacological sectors [33–36]. An effective technique to boost antioxidant activity of gold nanoparticles (AuNPs) is immobilization, capping, or functionalization of antioxidant ligands onto their surface. It significantly improves the reactivity of the functional groups of coated ligands. It means that assembling antioxidant ligands onto AuNPs has the potential to significantly improve antioxidant function.

Qian and his co-workers have shown a new nanocomposite-based technique for appraising H_2O_2 foraging activity based on H_2O_2-facilitated SiO_2/AuNP growth. The consistent change in the plasmon absorption band is well correlated with antioxidants' ability to scavenge H_2O_2 [37]. The powerful and tunable optical signal produced by the formation and the growth of AuNPs (1–100 nm in diameter), having characteristic surface plasmon resonance band in the visible range of 500–600 nm, could be used to fabricate biosensors with high reliability [38, 39]. Scampicchio's group has focused on polyphenol-mediated growth of AuNPs and introduced a nanoparticle-based method for assessing the antioxidant potential of phenolic acids found in edibles [40]. Optical properties of polyphenol coated nanoparticles have proven to be a good device to co-relate with the antioxidant properties of the phenolic acids involved. This relationship is anticipated because these compounds' antioxidant properties and ability to reduce Au(III) both show their tendency to bestow electrons. The electrochemical studies became a powerful tool and make this method extremely useful for assessing the antioxidant potential of foods and beverages. Compounds with lower oxidation potentials exhibited higher antioxidant activities [41, 42]. Some researchers have noticed a change in the anodic peak potential of some organic acids, including propyl gallate (+0.21 V), caffeic acid (+0.35 V), protocatechuic acid (+0.47 V), ferulic acid (+0.47 V), and vanillic acid (+0.81 V). The same trend is observed with AuNPs with the maximum absorbance values by Scampicchio and his co-workers (Fig. 3.6 and Fig. 3.7) [40].

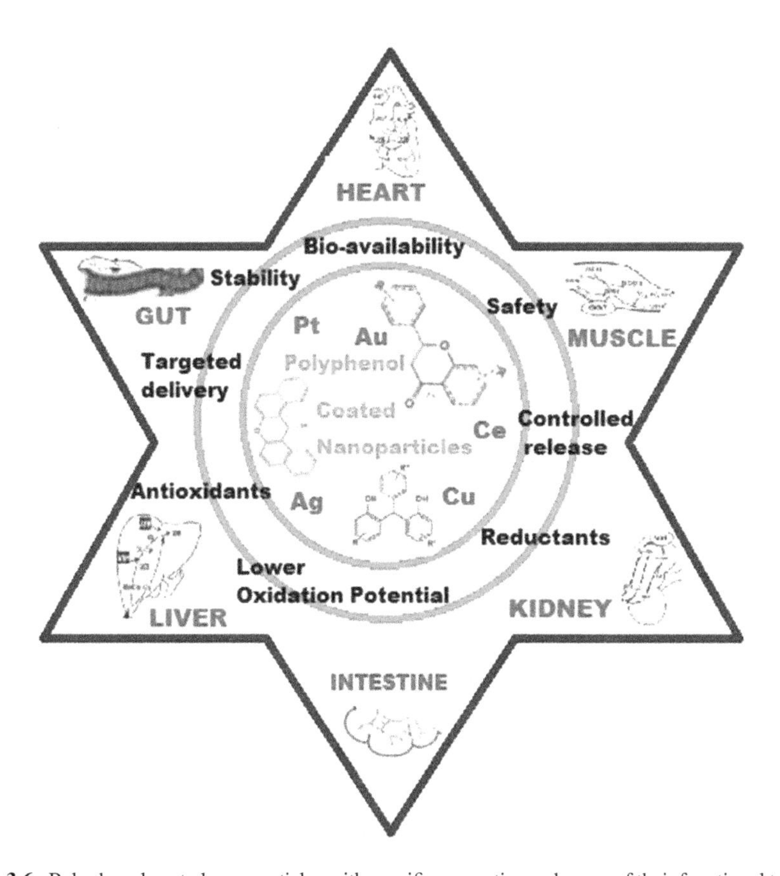

Fig. 3.6 Polyphenol coated nanoparticles with specific properties and some of their functional targets

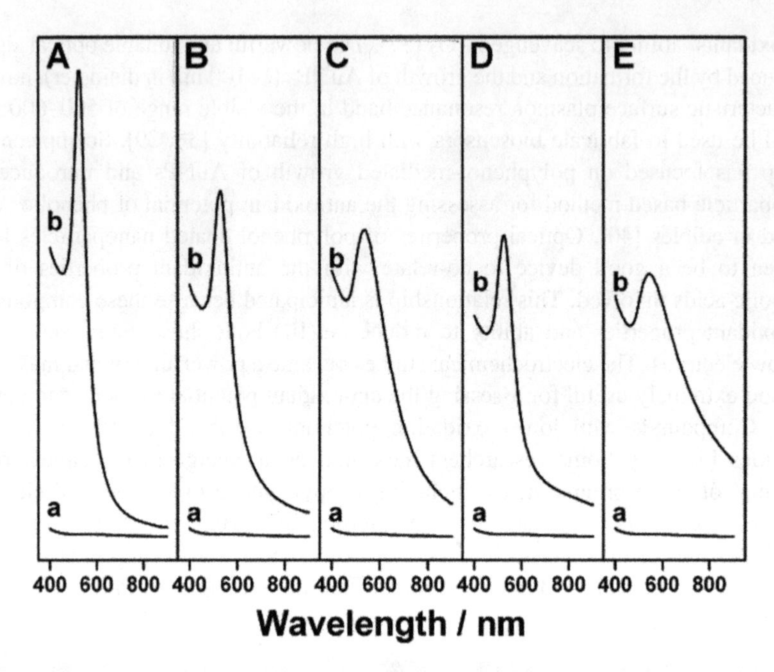

Fig. 3.7 Absorbance spectra of AuNPs formed in the (a) absence and (b) presence of 300 *μM concentrations of different phenolic acids:* (A) propyl gallate, (B) caffeic acid, (C) protocatechuic acid, (D) ferulic acid, and (E) vanillic acid [40]

The lowest anodic peak potential corresponds to the maximum ability for reducing Au(III) to AuNPs(0). The optical and electrochemical properties of phenolic acids is due to their chemical structure and particularly to the number of hydroxyl groups. Xia et al. has also found that the antioxidant property can be linked to the reducing power of the moieties. Reductones, considered to be having a high reducing strength, are thought to react not only with peroxides directly but also with certain precursors to prevent peroxide formation [43].

Nie et al. synthesized trolox (Au@Trolox) and chitosan coated gold nanoparticles with the anticipation that the production of novel self-assembled nano-sized antioxidants with multiple functions will be aided by the synthesis of water-soluble antioxidant-functionalized AuNPs. For the delivery of high-efficiency antioxidants to particular tissues or cells, these hydrophilic water-soluble AuNPs may again be functionalized with any receptor molecules for particular antibody–antigen or ligand–receptor interactions. In (Au@Trolox), AuNPs are coordinated to Trolox through Au-S bonding of thiol ligand. The antioxidant function of "Trolox," a water-soluble equivalent of Vitamin E, was proficiently improved by the self-assembly of chromanol rings of Trolox ligand onto the surface of AuNPs, as demonstrated in DPPH radical scavenging experiments. This organized and ordered vitamin E equivalent had an eight-fold higher radical scavenging capability when compared to the Trolox monomer [44]. In an another study, Liu et al. has developed, a gold nanoparticle-associated salvianic acid A (Au@PEG3SA) composite by a layer by layer self assembly process. This nanocomposite with salvianic acid A increased the rate of DPPH radical scavenging action by nine times compared to just bare salvianic acid monomer. Since there are only a lesser quantity of antioxidants within cell membranes, therefore, lipid peroxidation starts once the antioxidants get depleted. So to meet out the further requirements, exogenous antioxidants have to be supplemented to anti-lipid peroxidation defense. Salvianic acid is a type of amino acid that is found in plants.

In a similar way to Vitamin E, salvianic acid monomer inhibits lipid peroxidation, leading to a lipid radical-foraging reaction. Au@PEG3SA inhibits the propagation stage of lipid peroxidation more effectively than salvianic acid A monomer, resulting in a significant reduction in MDA levels in cells. As a result, the higher reaction rate constant of Au@PEG3SA as compared to the salvianic acid A monomer can account for the observed increase in antioxidant activity [45].

3.2 *Silver(Ag)*

Because of the unusual antimicrobial properties and biomedical applications, silver is the most widely used and researched noble metal among the coinage metals. Nowadays for the synthesis of silver nanoparticles green routes are specifically preferred. It has an additional advantage of getting provision of non-toxic and environmentally benign autocapped or functionalized biocompatible nanoparticles. The capping of phytochemicals such as flavonoids and lignins onto AgNPs shows higher degree of antioxidant potentiality with regard to scavenging action of free radical and reducing capacity. Number of plant and leaf extract mediated synthesis of AgNPs were done and the presence of these phytochemicals, via hydrogen atom transfer (HAT) and/or single electron transfer (SET) mechanisms, their free radical scavenging action and antioxidant capacities were assessed [46–50].

Some ornamental plant extracts are also studied which are able to reduce Ag^+ to Ag^0. The ingredients of the flowers act as reductants for Ag^+ as well as coating agents for phyto-synthesized silver nanoparticles (AgNPs). Studies have shown that the herbal green AgNPs (*Hyacinthus orientalis* L. and *Dianthus caryophyllus* L.) exhibited high values of antioxidant activity ranging between 88% and 97%. White carnation–AgNPs showed strongest antioxidant properties (97.38%). The DPPH free radical scavenging assay was used to evaluate the antioxidant activity of Ag nanoparticles, with vitamin C serving as a positive control. The results have shown that selenium, platinum, and silver nanoparticles have improved DPPH scavenging action [51, 47].

Delayed healing of wound is a major problem with diabetes patients. Number of studies have shown that augmented inflammation is caused by an increase in oxidative stress and a reduced antioxidant level. As a consequence, a large amount of ROS is produced, which causes inflammatory cells to apoptose prematurely. Recently, role of silver nanoparticle in diabetic wound healing therapy has been revealed by many researchers. Although the exact mechanisms of silver nanoparticles in bactericidal process in recovering from injury are unclear, a unique Ag + loaded zirconium phosphate nanoparticle has been found to play an important role in diabetic wound healing [52]. Recently, there are a few nanoparticles prepared in such a way that they may work as scavengers for free radicals. Amit Kumar Mittal et al. determined that the antioxidant role of AgNPs released by plant phenolic components was comparable to standard TROLOX using Rhododendron dauricum flower extract using the DPPH assay [53]. The capping of oxidized polyphenols or carboxyl protein was thought to be responsible for AgNPs' long-term stability. Mittal et al. have also fabrication bimetallic nanoparticles in the manner to extract their antioxidant efficacy. They synthesized bimetallic nanoparticles (Ag-Se) using quercetin and gallic acid as stabilizer. The antioxidant potential of monodispersed nano-antioxidants was assessed in vitro using azino-bis-ethyl benzthiazoline-sulfonic acid (ABTS), 3-(4,5-dimethylthiazol-2-yl)-2,5-diphenyltetrazolium bromide (MTT), and the DPPH assay. The nanocomposite's antioxidant efficiency was found to be between 59 and 62%. The redox potentials of the capping agents, namely quercetin and gallic acid, on the surface of Ag-Se nanocomposite ascertained the strong reducing nature of the effective antioxidant as H^+ donors and $^1O_2$ scavengers [54].

Sotiriou and his group demonstrated plasmon enhanced proton-coupled electron transfer (PCET) at near-IR wavelengths (700–1100 nm) using SiO_2-coated silver nanoparticles functionalized with gallic acid showed plasmon enhanced proton-coupled electron transfer (PCET). As the bond dissociation energy (BDE) of the phenolic OH group of gallic acid decreases, laser excitation incited the rate of hydrogen atom transfer (HAT) from gallic acid to DPPH radicals, accelerating the DPPH scavenging kinetics [55].

3.3 *Ceria(CeO_2)/Yttria(Y_2O_3)*

Nowadays, Ceria (CeO_2) nanoparticles, an oxide of first member of rare-earth element in the lanthanide series, are commonly used for the absorption of ultraviolet radiation. Nanoceria has a number of benefits over other experimented antioxidant approaches when used for medicinal purposes, e.g., the delivery of nano-encapsulated antioxidant enzymes like SOD or catalase has the drawback of being selective to just one form of reactive oxygen species present that each enzyme can scavenge, while numerous factors are involved in neurodegenerative diseases. In that gloomy situation, nanoceria has been shown to reduce ROS concentration. Furthermore, the amount of enzyme that is to be distributed inside the nanostructure can be minimized. Nanoceria, on the other hand, has both catalase and SOD-mimetic activity that is redox state dependent, and the catalysis of the SOD-mimetic enzyme is superior to that of the endogenous enzymes [56–59]. Seal and Self group have also observed that the cerium oxide has the ability to imitate the properties of SOD, which is an endogeno cellular protection against free radicals [60]. Das and his group have used ceria nanoparticles as oxygen sensing [61], while Yu et al. used it as an automotive catalytic converter [62]. Reports revealed that due to its excellent potentials to scavenge free radicals, cerium oxide (CeO_2, ceria) plays a significant dynamic role. CeO_2 nanoparticles are currently being studied for their potential use in quenching ROS in biological systems. CeO_2 nanoparticles have been shown to provide neuronal benefits in various researches [63] and ocular protection [64]. Niu et al. in a recent study have found that CeO_2 nanoparticles have potential to protect the heart from oxidative and inflammatory damages caused by monocyte chemotactic protein-1 expressions in the heart [65]. Colon et al. have observed that cerium oxide nanoparticles reduce reactive oxygen species (ROS) and control SOD, thus protecting the gastrointestinal epithelium from radiation-induced injuries [66]. These studies have shown that cerium oxide nanoparticles act as a potent antioxidant and a unique defensive agent in guarding cells and tissues by scavenging free radicals in a regenerative manner to cause any damage.

Another oxide of this category, yttrium oxide is also being considered as a significant antioxidant compound. Since it has having the highest oxidation potential or free energy for formation of oxides from elemental yttrium as compared to any known metal oxide under normal temperature and pressure, it can be distinguished even by minor deviations from its regular stoichiometry by the atmospheric absorption of H_2O and CO_2. Monoclinic B form of yttrium oxide is more or less similar to its particular hexagonal polymorph close-packing A form (hcp).

It has six coordination states in contrast to the A type, that has seven coordination states. The yttrium cations are in non-equivalent locations in the crystal of B type structure, which is significantly less dense than the A form version [67–70]. Since yttrium behaves like lanthanide elements, similar to cerium, and its cationic radius is in the same range as the lanthanides, Y_2O_3 and CeO nanoparticles display identical biological activities. Even though yttrium has no f-orbital electrons in its stable state, its cation has an inert gas configuration.

Non-stoichiometry is observed in both ceria and yttria, but the degree of non-stoichiometry in ceria is much greater reflecting the possession of inert gas configuration in yttria cation. Cerium oxide is found as monodispersed particles having single crystals and little twin interfaces, as well as a high lattice parameter. It has a cubic fluorite structure, while yttria has a cubical structure [71, 72]. Due to this non-stoichiometry of ceria, the cerium atoms are characterized by both +IV and + III oxidation states. Recent studies using X-ray photoelectron spectroscopy and near edge X-ray absorption spectroscopy have shown that as particle size decreases, the concentration of Ce^{3+} relative to Ce^{4+} increases, with a minimum 6% concentration of Ce^{3+} in 6 nm nanoparticles and 1% in 10 nm particle size [73]. These nanoparticles are known to function as strong and recyclable ROS scavengers by shuttling between Ce^{3+} and Ce^{4+} oxidation states [74]. The double oxidation states clearly indicate the presence of oxygen vacancies or defects in these nanoparticles. It means that the existence of dual oxidation states (Ce^{3+} and Ce^{4+}) in the structure of ceria might lead to a self-reformative redox cycle between Ce^{3+} and Ce^{4+}, giving rise to the formation of oxygen defects or oxygen vacancies on their surfaces, providing a large number of active sites for the scavenging action over free radical [75]. Ceria nanoparticles exhibit redox chemistry as a result of this property, making them a striking moiety for catalytic applications as well. The ability of cerium and yttrium oxide nanoparticles to protect cells from their cessations induced by oxidative stress seems to be reliant on the particle's structure and independent of their size at the range of 6–1000 nm. This may be beneficial in the healing of wounds in diabetic patients. Neutrophils and macrophages are largely unaffected by these nanoparticles, and they protect cells from oxidative stress deaths. The conventional antioxidant properties of the nanoparticles are responsible for this defense mechanism.

Nanoparticles of other metal oxides for their similar radical scavenger behavior were also deliberated, such as aluminum oxide (Al_2O_3), which is commonly known as alumina. Aluminum or alumina are considered to be common contaminants of food and water. Though their cytotoxicity and genotoxicity were investigated by many researchers, Radziun et al. have found that there was no significant increase in apoptosis or decrease in cell viability suggesting that aluminum oxide nanoparticles at particular concentrations has no cytotoxic effects on the selected mammalian cells [76].

It differs in crystal structure from ceria and yttria, but it is thermodynamically very stable at all temperatures. The oxygen atoms occupy hexagonal close-packing, while Al^{3+} ions occupy two-thirds of the octahedral sites in the lattice, giving it a corundum-like structure. Yttria, on the other hand, has a cubical structure as mentioned above [72].

Becker et al. have cited three different theories for cerium oxide and yttrium oxide particles' oxidative stress-protective behavior. Initially, they may act as direct antioxidants, inhibiting the formation of reactive oxygen species (ROS). Then they can directly trigger a low level of ROS development, which rapidly induces a ROS defense system before any exogen moiety-induced cell death course is completed. This can be achieved at preparatory level that occurs when cells are exposed to particulate matter that produces low levels of reactive oxygen species (ROS) [77]. Schubert et al. [78] have also shown that nanoparticles made of cerium and yttrium oxides are comparatively nontoxic to cultured cells and capable to protect HT22 nerve cells from oxidative stress induced by exogen glutamic acid. Based on these findings, it's possible that ceria, yttria, and alumina may be beneficial in the curing of many diseases, including diabetic wound healing therapy mediated by ROS scavenging potentials. Furthermore, for providing innovative solutions for support to long-duration human space exploration and applications to health issues, Gianni Ciofani and group [79] of Italian Institute of Technology provided information which is maintained in *database by the ISS (International space station) Program Science Office. It is* developed from the findings that the long-term exposures to microgravity causes severe alterations in skeletal muscle, and one of the reasons for these deleterious effects may be related to oxidative stress in muscle cells as a result of excessive production of reactive oxygen species (ROS). ROS overproduction can be juxtaposed with supplemental antioxidants consumption. Since traditional supplements have limited activity over time and need repeated administrations. Cerium oxide (CeO_2) nanoparticles as a novel agent, was identified with prolonged ROS scavenging activity. This ROS scavenger operates in a self-regenerating redox cycle, thus not needing repeated administrations. This explains the nanoceria-based countermeasures to microgravity-induced oxidative stress. To make them biocompatible, a team led by Rice chemist Vicki Colvin [80] has developed cerium oxide spheres that are compact and uniform and coated with a thin film of fatty oleic acid. They claim that their research will help patients with harrowing brain injury, heart arrest, and Alzheimer's disease, as well as cancer patients suffering from radiation-induced side effects. Cerium oxide nanocrystals have the ability to absorb and release oxygen ions by redox reactions, which could safeguard astronauts from long-term exposure to radiation in space and by reducing the effects of aging and even helping to slow down the aging process. It is the same process that allows catalytic converters in cars to absorb and eliminate pollutants. These particles are small enough to be injected into the bloodstream when organs need protection from oxidation, particularly after traumatic injuries, when damaging ROS increases dramatically. The cerium particles go to work immediately, absorbing ROS free radicals, and the process continues as the particles have potential to revert back to their initial state. The oxygen species released in the process. As cerium oxide cycles between cerium oxide (III) and cerium oxide (IV), it remains relatively stable.

The nanoparticles in the first state have holes at their surface that function like a sponge, absorbing oxygen ions. Cerium oxide (III) catalyzes the reaction that essentially deactivates free radicals by trapping oxygen atoms and converting back to cerium oxide (IV). Cerium (IV) oxide particles gradually liberate the oxygen they have taken in and return to cerium(III) oxide, allowing them to break down free radicals over and over again repeatedly. The polymer coated 3.8-nanometer sphered nanoparticles effectively act as scavengers of oxygen. A thinly coated nanoparticle has been shown to be fairly permeable to oxygen while still being strong enough to withstand several cycles of ROS suction. The researchers discovered that their most powerful cerium(III) oxide nanoparticles performed nine times better than a typical antioxidant, Trolox, at first exposure and sustained well through 20 redox cycles when tested with hydrogen peroxide, a strong oxidizing agent [81]. Smaller sized nanoparticles containing enough Ce(III) were found to be more scavenging towards H_2O_2. Moreover, surface coating does not impose any restriction on the reaction between the surface of the Ce(III) and H_2O_2.

3.4 *Copper/Copper Oxide (CuO)*

Subbaiya et al. have synthesized copper nanoparticles (CuNPs) from plant source and analyzed its antioxidant activity by the FRAP and hydrogen peroxide scavenging assay [82].

Din et al. have found that the concentration of plant extract plays a dominant role in reducing and stabilizing the CuNPs. The concentration of phytochemicals gets raised as the concentration of plant extract is improved, and the reduction of copper salt is enhanced as well. Smaller sized nanoparticles were obtained as a result of the metal salt's rapid reduction [83].

With the π electrons and carbonyl groups, some flavonoids can chelate with the CuNPs.

The existence of hydroxyls and carbonyl functional groups in flavonoids as in quercetin and santin resulted in substantial chelating action.

These groups were found to chelate with copper nanoparticles. Numbers of researches have shown that different polyphenolic phytochemicals as quercetin, catechins, flavanones, isoflavones, santin, penduletin, alizarin, pinocembrin, anthocyanins, flavones, tannins, and saponins, which are assumed to have antioxidant protein molecules (superoxide dismutase, catalase, glutathione), have a substantial reducing activity in chelation, formation, and stabilization of copper nanoparticles,

Flavonoids contain many polyphenolic compounds such as ellagic acid, gallic acid, phenylpropanoids (phenylalanine, tyrosine), terpenoids, cysteine proteases, curcuminanilineazomethine, ascorbic acid, eugenol, and alkaloids (discussed in previous chapter), which serve as reducing agents. During the tautomeric conversions of the enol form to the keto form, the reactive hydrogen atom in the flavonoids is most likely released, and reduce copper ions to form copper nuclei for the growth of CuNPs. Das et al. have explored the antioxidant and antibacterial activity of copper oxide (CuO) nanoparticles and reported that CuO nanoparticles show free radical scavenging capacity up to 85% in 1 h is relatively higher in comparison to other metal oxide nanoparticles [84].

Ghosh et al. [85] have done biogenic synthesis of monodispersed CuNPs using medicinal plants. The conversion of Cu^{2+} to CuNPs is indicated by the development of reddish brown color. Literature supported the surface plasmon peak of CuNPs at around 560 nm. Ghosh group showed that both porcine pancreatic α-amylase (PPA) and unprocessed murine pancreatic and intestinal amylase were inhibited by biogenic CuNPs. They demonstrated that the CuNPs have antidiabetic properties. In the giant freshwater prawn *Macrobrachium rosenbergii,* upraised concentrations of Cu^{2+} ions result in substantial inhibition of the digestive enzyme amylase, as well as an increase in the activity of antioxidant enzymes superoxide dismutase (SOD) and glutathione-S-transferase (GST). Fluorescence and circular dichroism were used to validate the interactions between CuNPs and PPA. Spectral changes indicated the alterations in the native conformation of the protein structure. Changes in the native conformation of the protein structure were suggested by spectral indications [86–88]. In vitro experiments have shown that the Cu (II) ion and its complexes have high α-glucosidase curbing activity, much greater than the clinically used acarbose. It has been documented that hyperglycemia related with prolonged diabetes produces reactive oxygen species (ROS), which causes oxidative stress, which is a major contributor to lipid peroxidation and membrane disruption [89]. Two effective line of treatment for diabetes prevention are the inhibition of oxidative damage with free radical scavengers and slowing down the action of digestive enzymes such as α-amylase and α-glucosidase. *Dioscorea bulbifera* tuber extract reduced CuNPs exhibited superior antioxidant activity, may contribute in the prevention of oxidative stress, a major factor in the advancement of diabetes and its associated physiopathological disorders.

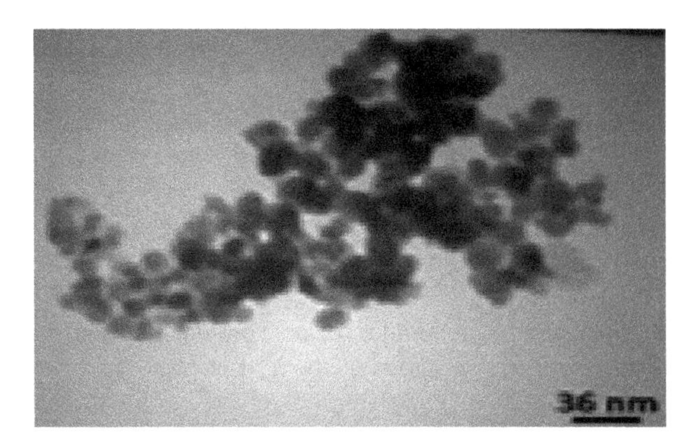

Fig. 3.8 Transmission electron microscopy (TEM) image of TiO_2 nanoparticles

Literature also recommends the baked copper "Tamra Bhasma" has a high antioxidant capacity and therefore has hepato-protective efficiencies. The usage of "Tamra Bhasma" for free radical scavenging action has recently been documented in pharmacological studies [90, 91]. Microelement concentrations have been shown to aid wound healing in former studies.

In the meantime it is worthwhile to mention that wound healing problems have been identified in patients with diabetes-related hyperglycemia and copper deficiency. Many studies have supported the use of "d" block transition metal nanoparticles, specifically CuNPs, as curative agents for T2DM management. It gives substantial proof for biogenically synthesized CuNPs' antidiabetic ability and suggests that they may be beneficial as preventive agents against the onset and progression of free radical-induced diabetes mellitus and its aggravations [92–94] (Fig. 3.8).

3.5 *Titanium Dioxide (TiO_2) Nanoparticles*

Hatami et al. have observed the antioxidant enzymes activities of superoxide dismutase (SOD), peroxidase (POX), and catalase (CAT) in *Hyoscyamus niger* L. using nanoscale level titanium dioxide particles (NT) and bulk titanium dioxide (BT). According to the findings, SOD activity increased with raising titanium dioxide concentration in both nanoparticles and bulk treated plants, Plants treated with 40 mg/L NT have shown the highest and lowest POX activity, respectively.

In general, they discovered that the activities of all enzymes tested were higher in NT-treated plants than in BT-treated plants, with the exception of CAT activity [95, 96].

3.6 Platinum Nanoparticles

For biological applications the prima requisite for nanomaterials is its biocompatibility. Platinum nanoparticles (PtNPs) have a high catalytic activity, as well as the capability to reduce intracellular reactive oxygen species (ROS) levels and inhibit provocating pathways. Deborah et al. studied the enzymatic activity of PtNPs, and analyzed their ability to neutralize reactive oxygen species (ROS) produced by extreme disorders within cells [97].

Since polyphenols and associated biologically active substances have been investigated as effective antioxidants. Yi Liu et al. have attempted to comprehend the possible effects of Pt NPs on polyphenol antioxidant activity. By oxidizing polyphenols to their congruous o-quinones, platinum nanoparticles (Pt NPs) were discovered to have catechol oxidase-like action. Their line of study was based on four approaches, through which they confirmed the said activity by Pt NPs. They started with the monitoring of the oxidation reaction of polyphenols catalyzed by Pt NPs using UV–Vis spectroscopy. Ultrahigh-performance liquid chromatography separation is used to identify the oxidized products of polyphenols, which is accompanied by high-resolution mass spectrometry in the second phase. Then in the third step the consumption of O_2 during the oxidation reaction is confirmed with electron spin resonance (ESR) oximetry techniques. Lastly in the fourth step with ESR using spin stabilization, the intermediate products of semiquinone radicals are determined which are formed during the oxidation of polyphenols [98]. Chen et al. have used electron spin resonance spectroscopy to compare and contrast the results of Au, Ag, and Pt nanoparticles (NPs) on the antioxidant activity of vitamin C (sodium salt of L-ascorbic acid, NaA) in the context of studying the interactions between noble metals and other chemical components of consumer goods. Their chemical experiments showed that Pt NPs have ascorbate oxidase-like activity, oxidizing NaA, while Au and Ag NPs did not. Because of the ascorbate oxidase-mimetic activity of Pt NPs, there was a significant loss of antioxidant activity of NaA for scavenging hydroxyl and superoxide radicals. Pt NPs' ascorbate oxidase-mimetic activity, on the other hand, is unconditionally dependent on particle size. They also discovered that Pt NPs with ascorbate oxidase-simulative activity inhibit the cytoprotective effect of NaA on cells that have been exposed to oxidative stress in their in vitro cell studies [99].

The free radical scavenging behavior of polyacrylic acid (PAA)-capped platinum nanoparticle species (PAA-Pt) formed by alcohol reduction of hexachloroplatinate was investigated by Watanabe and his team. The residual peroxyl radical produced from 2,2-azobis (2-aminopropane) dihydrochloride (AAPH) by thermal decomposition in the presence of O2 was measured using electron spin resonance, and the residual 1,1-diphenyl-2-picrylhydrazyl (DPPH) radical was measured spectrophotometrically. PAA-Pt scavenging activity on these two radicals has been studied in a dose-dependent manner. The pivotal factor was platinum. They discovered that PAA-Pt inhibited the propagation of linolate peroxidation by lowering the rate of oxygen consumption needed for the linoleic acid peroxidation chain initiated by

AAPH. Dose-dependent inhibition of the linolate peroxidation by PAA-Pt was also confirmed by the thiobarbituric acid test [100].

Martin et al. used Fenton-treated HO-Diamond nanoparticles as a support for gold and platinum nanoparticles with sizes approaching 2 nm. They discovered that the materials generated (Au/HO-DNP and Pt/HO-DNP) have strong antioxidant activity against reactive oxygen species induced in a hepatoma cell line, indicate good biocompatibility. These nanocomposites exhibited about a two-fold higher antioxidant activity than glutathione. But Au/HO-DNP nanoparticles have been proven to be most active material against cellular oxidative stress [101].

Alloy-structured bimetallic nanoparticles consisting of gold and platinum were prepared by Kajita et al. through citrate reduction method and using a complementary stabilizing agent pectin (CP-Au/Pt). The % mole ratio of platinum was varied from 0 to 100% and well dispersed in water. CP-Au/Pt consisting of more than 50% platinum successfully quenched hydrogen peroxide (H_2O_2), while any aliquot of CP-Au/Pt was able to quench superoxide anion radical. The quenching action of these two reactive oxygen species by CP-Au/Pt is perceived in dose-dependent manners. The CP-Pt is found to be the strongest quencher. As catalase, CP-Pt generated O_2 by the decomposition of H_2O_2. The quenching action of CP-Pt was actually verified by a superoxide dismutase assay kit. This quenching activity persisted as SOD also. Kajita suggested that CP-Pt may act as a SOD/catalase mimetic when taken in a composite manner [102]. Reactive oxygen species (ROS) production in tumor cells is found to be excessive, which promotes angiogenesis, metastasis, and secondary tumor development by increasing the number of receptors on the surface of target to certain growth factors and matrix metalloproteinases [103]. Antioxidants are the blissful gift from the nature as cancer chemo-preventive agent that can check, prevent, or suppress carcinogenic advancement. Ghosh et al. have discovered that nanoparticles are not only cytotoxic towards cancer cells, but they also serve as antioxidants. They discovered that mixed bimetallic Pt–PdNPs had a higher antioxidant capacity than individual PtNPs and PdNPs, scavenging 38.49 percent of the DPPH radical and exhibiting potent anticancer and antioxidant activity. In the similar pattern, when compared to individual PtNPs or PdNPs, a synergistic enhancement of scavenging activity against superoxide radical, nitric oxide, and hydroxyl radical scavenging was also observed with Pt–PdNPs. Chemically produced nanoparticles, on the other hand, had a lower free radical scavenging activity [104]. It was reported recently that in Japan, since the last 60 years or more, a mixture of Pd and Pt nanoparticles were being used to treat chronic diseases, with remarkable superoxide dismutase and catalase activity in an in vivo mouse model, proving to be a powerful tool to monitor age-related skin disease caused by oxidative impairments [105]. Team of Kim has provided an another credible information about PtNPs, that displays superoxide dismutase/catalase mimics with important anti-aging properties and extending life expectancy in *Caenorhabditis elegans* [106]. With these observations, usage of Pt as a constituent could be a useful technique for designing and exploring the impact of combination of bimetallic nanomaterials.

3.7 *Iron/Iron Oxide Nanoparticles*

MRI, drug delivery systems, hyperthermia, immunoassay, and tissue repair and detoxification are only a few of the applications of superparamagnetic iron oxide. Via magnetic administering, magnetic nanoparticles loaded with antioxidant enzymes such as superoxide dismutase (SOD) or catalase (CAT) have been used in drug delivery systems. High concentrations of antioxidants can be therapeutically directed to particular locations with enhanced levels of ROS using magnetically sensitive antioxidant nanocarriers.

Magnetic nanoparticles containing catalase demonstrated rapid cellular absorption and improved resistance to oxidative stress damage caused by hydrogen peroxide. Magnetic nanoparticles have been used for targeted enzyme remediation, which may be used to treat oxidative damage-related cardiovascular diseases. According to recent research, iron oxide nanoparticles have antioxidant properties, and their activity increases as particle size decreases. Czerniak et al. have used iron oxide nanoparticle (IONP) and using the ferric reducing antioxidant potential (FRAP) and 2,2′-diphenyl-1-picrylhydrazyl (DPPH) methods, the antioxidant capacities of acetonic, ethanolic, and methanolic extracts of rapeseed oils were determined successfully. They used sinapic acid (SA), caffeic acid (CA), gallic acid (GA), ferulic acid (FA), vanillic acid (VA), and trolox (TE) as standard antioxidant solutions to validate their findings [107].

Bishnu Prasad and his co-workers researched the antioxidant and antimicrobial properties of Himalayan honey (HH) loaded onto the iron oxide nanoparticles (IO-NPs). The Himalayan honey was collected from high altitude regions because it contained substantially more antioxidants than honey from lower altitudes. Such researches prove to be useful for effective designing, synthesis, protection, and encapsulation of nanoparticles based on natural product and their controlled release to bioactive molecules. Honey plays role of a surfactant. DPPH free radical was used to test the antioxidant activities of bare HH, IO-NPs, and honey loaded IO-NPs. In their dose-dependent experiment, they found two-fold increase in the antioxidant activities of IO-NPs after being loaded with HH as compared with honey alone.

The enhancement of the antioxidant activity of phenolics and flavonoids found in honey can be accredited to the IO-NPs that increases the antioxidant activity manifold. The Fe_3O_4 nanoparticles also display good antioxidant activity by considerably scavenging the free radicals, may be due to the electron transfer from the Fe^{2+}/Fe^{3+}systems of IO-NPs. Honey loaded IO-NPs shown an enhanced radical scavenging activity upto 80% [108]. Bhattacharya et al. also supported the work by achieving 89% inhibition of DPPH radicals with Fe_2O_3/C nanocomposites [7].

At room temperature, Narendhar et al synthesized iron oxide, iron-cobalt core shell nanocomposite, and zero valent iron nanoparticles using gallic acid as the surfactant and reducing agent, followed by co-precipitation, micro emulsion, and wet chemical techniques. Using the DPPH assay, the antioxidant capacities of the samples were compared to BHA, BHT, rutin, quercetin, and the blank. With increasing concentrations, the chelation activity of iron oxide and zero valent iron samples was found to be the same. The free radical scavenging experiments were done with the same data. They discovered a decline in the activity, which they attributed to the relative inhibition of high concentrations of contents present in the media as well as the lowering of nanoparticle surface area due to increase in size on crystal growth. With the increase in the size of the nanoparticles, the most likely cause of the decline in free radical scavenging and antioxidant activity was thought to be a decrease in the surface area to volume ratio [109].

Sastry et al. have observed that the surface-functionalized iron oxide nanoparticles synthesized using the seed coat extract of *B. flabellifer* have shown significant scavenging activity against free radicals [110]. A team of Sandhya and Kalaiselvam also worked in the same pattern and synthesized iron nanoparticles. The formation of strongly crystalline inverse spinel magnetite iron oxide nanoparticles with an average size of 35 nm was confirmed by X-ray diffraction, and the UV–Visible absorption spectrum revealed a characteristic iron oxide nanoparticle peak at 352 nm. They observed that synthesized nanoparticles had strong radical scavenging activity against DPPH, hydrogen peroxide, and hydroxyl radicals, and they accredited this to the presence of polyphenols in the extract, which improved the nanoparticles' antioxidant properties [111].

Under green synthesis impetuses, many researchers have synthesized iron-oxide magnetic nanoparticles (Fe_2O_3NPs) and using different methodologies, their antioxidant properties were also evaluated. Surface functionalization with various natural antioxidants such as gallic acid, poly(Gallic acid), and curcumin in magnetic-silk core-shell nanoparticles were done efficaciously. Thus obtained Fe_2O_3 nanoparticle composites were found stable and showed better dispersion ability and effectual antioxidant properties. As compared to the non-functionalized Fe_2O_3 nanoparticles, surface-functionalized Fe_2O_3 nanoparticles with gallic acid exhibited 2- to four-fold higher values with DPPH antioxidant assay. The synergistic effect of Fe_2O_3 nanoparticles and gallic acid is responsible for the increased free radical scavenging activity.

The mechanism occurs through transfer of electron from Fe_2O_3 nanoparticle coated gallic acid to free radicals located at the DPPH's central nitrogen atom [112].

In addition to hemocompatibility and blood cell viability experiments, Szekeres et al. polymerized magnetite nanoparticles coated with gallic acid-shell in a reagent-free process and checked for antioxidant potential in the presence of H_2O_2 as ROS in situ at the surface of the particles in soft Jurkat cells. They observed a major reduction in oxidative stress caused by H_2O_2. In vitro tests have revealed that PGA@MNPs are both biocompatible and biologically active [113]. Generally, the nanoparticles synthesized through green routes have an additional property displaying radical scavenging activities which may be attributed to the extract's presence of bioactive components with strong antioxidant properties. As mentioned in Chap. 2, the phytochemicals such as flavonoids, tannins, saponins, and glycosides are all known to be found in the extract and have anti-free radical properties due to the presence of polyphenolic content inherently in their structures. Therefore, iron oxide nanoparticles synthesized through green pathway may prove to a candidate with substantiate potential and can be used in a variety of biomedical applications because of their high cyto-consonance and antioxidant actions.

3.8 Zinc Oxide Nanoparticles

Zinc oxide nanoparticles (ZnO NPs) have been one of the most widely used metal oxide nanoparticles in the last two decades, not only in industrial products such as rubber, paint, coatings, and cosmetics, but also in consumer goods and emerged as a promising material having excellent potential in biomedicines and in biological applications due to its exceptional biocompatibility, low toxicity and is economical too. In the field of anticancer, antidiabetic, antibacterial, an antioxidant ZnO NPs have released zinc ions as a result of their potent ability to reduce excess reactive oxygen species (ROS) formation and scavenge the system and additionally, helps in keeping the structural integrity of insulin for the effective treatment of diabetics. That's why the US Food and Drug Administration (FDA) is graded ZnO as a "GRAS" (generally recognized as safe) substance [114].

Zn is a significant component of over 300 enzymes that act in the cell membrane and maintain its consistency and involved in various aspects of cellular metabolism, including protein, lipid, and carbohydrate metabolism [115, 116].

Stan et al. have performed antioxidant activity of powdered ZnO NPs by modified DPPH method, in which they dispersed ZnO NPs in methanol mediated DPPH in a glass vial. Gradually deep violet colored DPPH radical solution in the presence of ZnO NPs became colorless or the color changed to pale yellow. The surface reaction was boosted by mild magnetic stirring. The supernatant was analyzed by EPR spectroscopy after every 30 min for evaluating antioxidant activity of ZnO NPs by its scavenging effect. On examining the experimental data, in contrast to chemically synthesized ZnO NPs, the radical purgation capacity of ZnO NPs prepared with plant extracts was found to be higher. This supports the earlier mentioned fact that phytochemical capping enhances the antioxidant activities [117].

Pavan et al. observed violet, green, and red luminescence at room temperature on exciting the ZnO nanoparticles at 378 nm, which is accredited due to oxygen vacancy and ZnO interstitial defects. These nanoparticles by scavenging DPPH radicals show good antioxidant activity [118].

Esen evaluated antioxidant and anti-radicalic activities of ZnO NPs. He determined antioxidant and radical scavenging activities by using BHA and a-tocopherol as standard. The various methods he followed were the ferric cyanate mediated reduction process of Fe^{3+} to Fe^{2+} method, reduction of Cu^{2+} ions with cupric method, measuring Fe^{3+} reducing abilities by the FRAP process, chelating activity of Fe^{2+} ions, radical scavenging activity of $ABTS^{\bullet+}$ species, and superoxide anion radical ($O2^{\bullet-}$) removal activity. He found that ZnO NPs are both effective at radical reduction and very efficient and effective at radical elimination. His results established that ZnO NPs have higher activity than the standard ones [119].

Shadmehri et al. while researching the antioxidant properties of ZnO NPs, graphene, and G-ZnO, researchers discovered that as the concentrations of ZnO NPs, graphene, and G-ZnO increased, the absorption of DPPH and ABTS radicals increased. Among them G-ZnO displayed highest order of their DPPH and ABTS scavenging activities while graphene the least. They attribute higher G-ZnO's antioxidant activity than ZnO NPs and graphene to synergetic effects between ZnO NPs and graphene, which can cause alterations in cytokines and intracellular calcium concentrations by producing reactive oxygen species and activating transcription factors [120]. When ethanol extract of *M. harentia* fruit was used to synthesize ZnO-NPs, then radical scavenging activity of that ZnO NPs was recorded maximum inhibitory action of 82.78 percent, suggesting that ZnO NPs have a high antioxidant activity [121].

Sharda Sundaram et al. have found the albino Wistar rats tissue homogenate resulted in rapid peroxidation on incubating it with ferric chloride alone. The biological tissues are attacked by the iron-induced peroxide formation. MDA (malondialdehyde) and other aldehydes were formed as a result of this process. In biological systems, aldehydes are normally reactive species capable of forming adducts and complexes. Aldehydes with TBA form pink colored chromogen complex with maximum absorbance peak at 532 nm. However, at a concentration of 600 mg/ml of nano-ZnO, the percent inhibition of lipid peroxidation as a function of varied concentrations of nano-ZnO was highest in the liver, followed by kidney tissues [122] (Fig. 3.9).

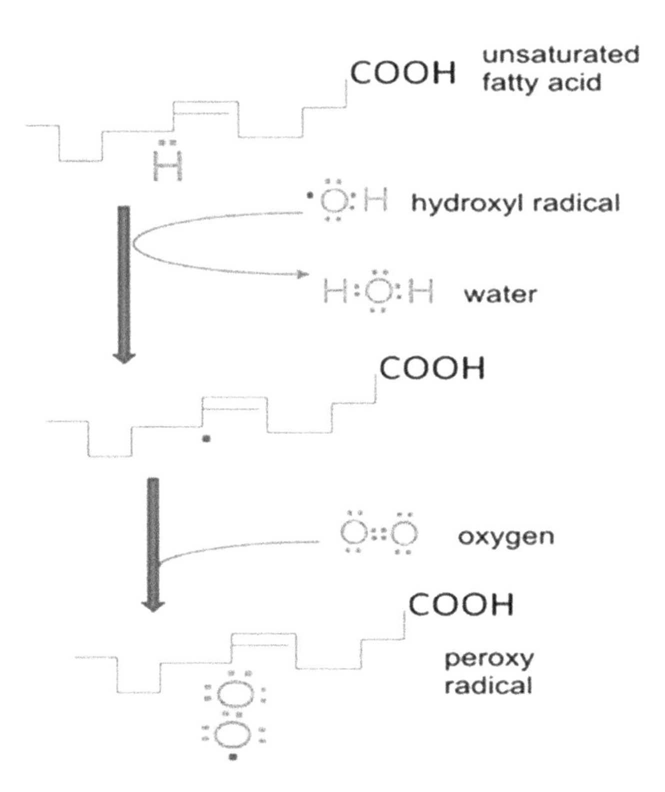

Fig. 3.9 Schematic representation of lipid peroxidation reaction [122]

While studying the phytostimulatory action of ZnO NPs on the germination and radical growth of Cicer arietinum (chick pea) seeds, Sharda et al. have been observed that the improved level of growth hormone "Indole acetic acid" is responsible for the faster growth rate of the radical sprouts and is solely due to the zinc-enriched ZnO NPs [123]. In another study, Sharda Sundaram et al. have investigated the impact of functionalized ZnO nanoparticles on the behavior of phytohormones in the *Solanum melongena* (brinjal) plant by measuring plant growth and development. The level of superoxide radical and hydrogen peroxide, lipid peroxidation, total phenolics, DPPH assay, SOD, POD, and GST activities were used to assess the potent antioxidant activity of functionalized nano-ZnO. At a specific concentration of nano-ZnO, it showed to have a positive effect on the plant [124].

According to Stehbens et al., by efficiently stimulating the cells to generate more SOD and CAT enzymes, the cells loaded with nano-ZnO had better performance and displayed higher antioxidant activities. Since SOD catalyzes the conversion of superoxide to peroxide and oxygen, the formation of superoxide and peroxide are intricately connected. As a result, it modulates ROS and is responsive to Zn [125]. Hydrogen peroxide's *cytotoxic* impact is checked by decomposing it to yield H_2O and O_2 by the enzyme CAT, localized in peroxisomes.

Another plant-dependent antioxidant study of ZnO nanoparticles has been done by López and her group. They have investigated the effects of zinc oxide nanoparticles on *Capsicum annuum* L. seed germination and seedling growth and the total phenols, flavonoids, and compressed tannins, as well as DPPH antioxidant potential were also calculated. Their findings showed that treating seeds with zinc oxide nanoparticles (ZnO NPs) increased seed germination within a week. They found a significant effect on radicle length and found that the seed vigor and seedling growth were influenced by zinc oxide nanoparticles, and the accumulation of required phenolic compounds in the radicle was encouraged [126].

4 Non-metallic Nano-antioxidants

4.1 Silica Nanoparticles as Nano-antioxidant

Silica, represented as SiO_2, at its nanoscale levels is one of the important contenders due to its chemical inertness, optical transparency, mechanical stability and above all biocompatibility for various diversified applications especially in medicine, pharmaceutical, and biomedical field. It is a wonderful moiety. Particle size, shape, porosity, crystallinity, and surface chemistry are all important factors to be considered for coating and functionalization.

Nanosized SiO_2 particles provide an excellent template for the immobilization of antioxidants onto it, yielding a value-added hybrid nanocomposites. Deligiannakis et al. functionalized commercially available SiO_2NPs by covalent grafting of natural antioxidants, gallic acid, which resulted in the different sized (8–30 nm) nanoparticles. The fabricated nano-antioxidants were evaluated for radical scavenging activity with

DPPH radicals. They observed two types of radical scavenging reaction mechanisms by SiO_2-GA nanoparticles, a secondary slow radical-radical combination reactions and a rapid H-atom transfer (HAT). SiO_2NPs, in general, opt quick HAT reactions [6].

Arriagada and group adsorbed Morin flavonoid ((2′,3, 4′,5,7-pentahydroxyflavone) on mesoporous silica nanoparticles (MSN) and evaluated its antioxidant potentials as a scavenger of hydroxyl radicals ($HO^{\cdot}$) and a quencher of singlet oxygen ($^1O_2$) [127]. From the structural point of view, having possession of ortho- and meta-hydroxyl groups in the catechol structure in the MSN covalently coated with phenolic antioxidants, they can easily donate hydrogen atoms to free radicals, thus preventing ROS formation and biological damages.

4.2 Fullerenes(C_{60})

Fullerenes are allotropes of carbon. *It* is a mixture of sp^2 and sp^3 hybridized carbon atoms in 5 and 6 membered fused ring system (Fig. 3.10). Their chemical properties are based on their bonding. It has carbon atoms with sp2 and sp3 hybridized valence shell. Since intermingled sp3 hybridization restricts free π-electron delocalization to some extent, it does not show super-aromaticity. But it possesses large amount of conjugated double bonds as having sp2 hybridized carbon atoms providing triply-degenerate low lying antibonding lowest unoccupied molecular orbitals showing considerable electron affinities by willingly ready to accept up to 6 electrons from any donor. Its geometry having delocalized πmolecular orbitals extending throughout the structure allows charge distribution over large volume which reduces the

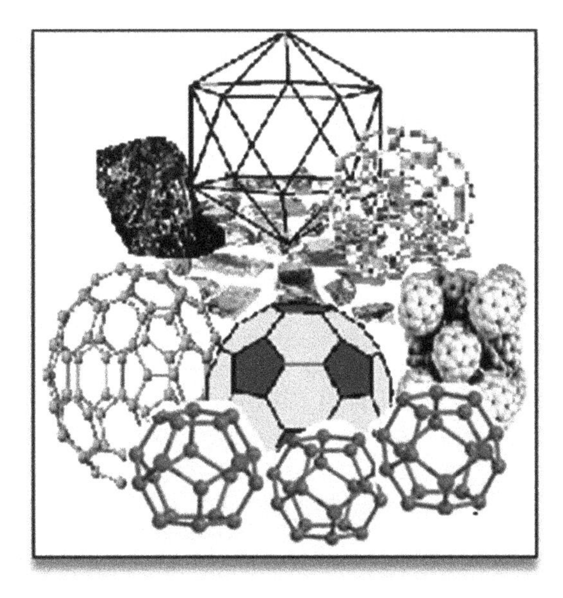

Fig. 3.10 A collage of different fullerenes

electrostatic repulsion at the donor site [128]. This partial delocalization of electrons due to conjugation and low lying LUMO make the basis for the fullerenes to act as antioxidants, promising a radical species attack extremely probable. According to reports, a single C_{60} molecule can be quenched by adding up to 34 methyl radicals onto it. It displays catalytic activity by reacting with a large number of superoxides that remained unconsumed. Owing to this peculiar property, fullerenes are mentioned as radical sponges, i.e., world's most efficient radical scavenger [129]. Fullerenes perform their electron/radical scavenging antioxidant activity by their ability to confine in the mitochondria and other organelles of the cell wherever the production of free radicals takes place causing diseases.

Nijla et al. in their work have revealed that without any polar organic solvents, drenched C_{60} suspensions have shown to be efficient and powerful liver-protective agent by protecting rodent's liver against free radical damage without any acute or subacute toxicity [130]. Intoxication of rats with carbon tetra chloride (CCl_4) resulted in the formation of trichloromethyl radicals (CCl_3•), an extremely soaring reactive species that quickly initiates the chain propagation reaction of lipid peroxidation, resulting in significant liver damages [131]. But no liver damages were observed in rats that had been pre-treated with C_{60} and then intoxicated with CCl_4 revealed the free radical scavenging ability of C_{60}.

A dose-dependent supply of pristine C60 showed a powerful liver-protective agent. These become water soluble on treating polyhydroxylated fullerenes ($C_{60}(OH)_{22}$) and C_{60} tris(malonic)acid that have polar groups, allow them to cross the cell membrane, especially through mitochondria, which is a source of cellular oxygen free radicals [132].

Nanoparticles may act as antioxidant or toxic, it is the dose that decides a drug will act as a medicine or poison. Igor et al. have worked on the impact of fullerene C_{60} on the *Escherichia coli*. They found that the results strongly depend on the number of factors, viz. experimental conditions, method of preparation, and solvents used for preparation as well as for an application. The measurement of oxidative stress induced by adding hydrogen peroxide and in the presence of fullerene C_{60} suspensions of various concentrations, its progression with time was documented. According to them, C_{60}'s capability to permeate through biological membranes, proton conduction, and deal with free radicals is the most probable reason of its defensive effect on E. coli, and it may thus be thought of as a mitochondrial harrowed antioxidant [133].

Many of the water-soluble fullerene derivatives such as fullerenol $\mathbf{C_{60}(OH)_{22}}$ have been recognized for their antioxidant properties. Kato et al. confirmed through beta-carotene bleaching assay that $C_{60}(OH)_{32}.8H_2O$ has antioxidant properties [134]. Cytoprotective action of water-soluble radical scavenger fullerene derivative, C_{60} @ poly(vinylpyrrolidone) against UVA irradiation was studied by Xiao and his co-workers [135]. Ultraviolet radiation A with a wavelength of 320–400 nm effects the human skin cells, causing the production of reactive oxygen leading to cell damage by instituting oxidative stress through catalytic dismutation of superoxide. C_{60} @ poly(vinylpyrrolidone) has shown the potential to penetrate deeper into the human skin epidermis because it is more resistant to oxidative crumbling than vitamin C and without photosensitization or cytotoxicity, it is possible to duck both UV skin injuries and skin aging. In the field of neurosciences also, fullerenols and malonic acid derivatives of C_{60}, which are water-soluble fullerenes, have drawn a lot of interest. As a very sensitive organ towards oxidative damages due to the triggering of aerobic metabolism of unsaturated fatty acids of brain, it has limited ability to regenerate the damages caused by the $O_2^{\bullet-}$ (superoxide), $^{\bullet}OH$ (hydroxyl) free radicals, and H_2O_2 molecules [136].

Fullerene products by scavenging intermediate peroxyl radicals show the ability to obstruct the lipid peroxidation chain reactions. On human embryonic lung fibroblasts, Sergeeva and co-workers tested two water-soluble fullerene derivatives, C_{60} and C_{70}, that contained solubilizing residues of amino acid and aromatic acid. They found that the chemical functionaries attached to the fullerene center, as well as the shape and size of the fullerene cage, have a major impact on ROS development. Therefore it is obvious that the chemical composition of fullerene C_{60} and C_{70} derivatives determines their antioxidant activity. The different cell responses to the derivatives are observed due to the chemical functionaries attached to the fullerene core as well as the shape and size of the C_{60} and C_{70} cages [137].

5 Polymeric Nano-antioxidants

There are so many polymeric nanoparticles and nanofibers serving promising role as nano-carriers and delivery vehicles. Using polymers for the fabrication of nanoparticles, it is of utmost important to remember that during entrapment or encapsulation, the nanoparticles should maintain its well-defined physicochemical structural properties, sensitivity, and compatibility with bioactive components. Only then the nano-antioxidants or drugs can be released in a controlled and predictable manner dodging the immune system before realization against external influences. Polymers used as nanocarriers for antioxidants should be biocompatible, biodegradable, wettability, and swelling property with high porosity and greater water holding capacity. Polymers of caprolactone (PCL), amino esters (PAEs), D,L-lactide (PLA) and lactic-co-glycolic acid (PLGA), vanillin oxalate (PVO) are such few examples of semi-crystalline biodegradable and biocompatible polymer nanoparticles with hydrolytically degradable crosslinking, low toxicity, physically tough making it a suitable contender for several drug delivery devices or serving as transporters for medicines, minerals, proteins, and other biomolecules such as peptides, DNA or RNA (Fig. 3.11).

Some examples of natural polymers that include proteins, polysaccharides, chitosans, curcumin, gum arabic, polypeptides, etc. are used for encapsulation of many antioxidants moieties to enhance their efficacy and efficiency to feat their natural characteristics such as bioavailability, biocompatibility, and biosafety making them superior over synthetic polymers. Though natural polymers have their inherent inconsistency which may adversely affect the encapsulated nanomaterials but due to bio-homogeneity encapsulated nano-antioxidants/nanomaterials get

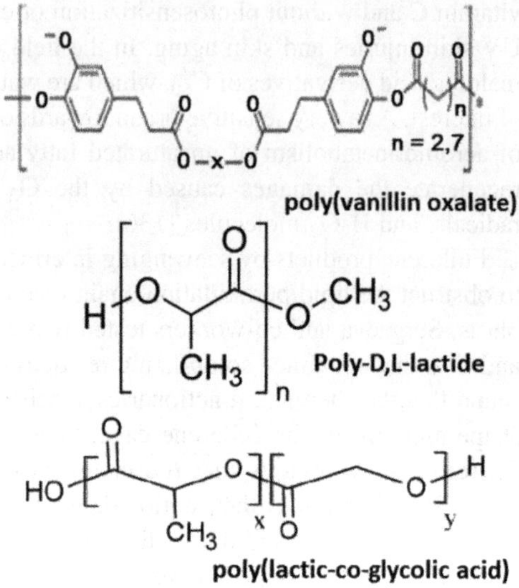

Fig. 3.11 Structures of some synthetic polymeric nanocarriers/delivery vehicles

poly(vanillin oxalate)

Poly-D,L-lactide

poly(lactic-co-glycolic acid)

released more rapidly than the synthetic polymers and their minimal side effects and conceiving conditions make them a preferred entity over synthetic polymers.

Some natural nano-antioxidants that are having polymeric structures are discussed in following sections:

5.1 Vanillin

Vanillin is a major component of natural vanilla and is an aromatic aldehyde (4-hydroxy-3-methoxybenzaldehyde) with a hydroxyl group para to aldehyde. It is commonly used as a flavoring agent in food, beverages, and cosmetics [138]. It was a new discovery to have strong antioxidant properties, preying superoxide and hydroxyl radicals and protecting membranes from protein and lipid peroxidation [139].

It is a general observation that vanillin readily gets decomposed in the upper digestive tract when taken orally or if injected intravenously, it quickly leaves the blood circulation. So it has limited or scanty therapeutic effects [140]. On the other hand, vanillin nanoparticles made of polymeric vanillin oxalate (PVO) act as a polymer pro-drug to improve the therapeutic effectiveness of vanillin by delivering it to diseased cells or tissues. PVO covalently combines with vanillin and releases it during its hydrolytic degradation, thereby showing a higher H_2O_2 preying activity than free vanillin and thus reducing H_2O_2-induced oxidative stress. Kwon et al. have reported that the PVO nanoparticles with dual stimuli response as therapeutic agents for diseases caused by oxidative stress. Kinetics of degradation by hydrolysis and release of vanillin of PVO is observed in a pH-dependent reaction. They found that at pH 5.4, the majority (80%) of vanillin was released only in 36 hours, but due to the acid-cleavable acetal linkages, only half the amount (40%) of vanillin was released at pH 7.4. Moreover, PVO nanoparticles displayed a highly efficient antioxidant and anti-inflammatory activities; PVO nanoparticles have shown a very high level of bio-adaptability at concentrations less than 500 _g/mL. In zest, PVO nanoparticles have been shown considerable therapeutic potential for oxidative stress-related inflammable conditions, due to H_2O_2 scavenging capability as it has H_2O_2 sensitive peroxalate ester bonds in its prop, which are believed to be spontaneously oxidized by H_2O_2 and break up into CO_2 [141].

5.2 *Curcumin*

Curcumin is a fat-soluble polyphenolic alkaloid compound having a wide range of valuable properties as anti-inflammatory, antimicrobial, anticarcinogenic, and antioxidant properties, i.e., having both biological and pharmaceutical characteristics. But its hydrophobicity, poor solubility in water, with poorer bioavailability in plasma and tissue and a shorter biological half-life period make it less accessible. So, to increase its bioaccessibility and antioxidant activity, it has to be treated with nanoparticles. When compared to isolated curcumin, nanoparticles encapsulated within curcumin shell shows many folds the antioxidant function in epithelial cells. The antioxidant function of the curcumin-loaded PLGA-based nanoparticle was investigated both cellularly and acellularly by Betbeder et al. The reason for the drastic enhancement in the antioxidant activity is ascribed to the efficient entry of curcumin facilitated by nanoparticle endocytosis into the air routes in epithelial cells. The researchers have observed that both curcumin-nanoparticle antioxidant and direct anti-nitrosant effects were found to be greater when curcumin was fabricated with PLGA-nanoparticle. The most probable reason lies in the nano-environment which supported the enhancement of antioxidant activity by providing proper concentration and curcumin interactions with ROS and RNS are facilitated [142].

Das and Das observed that curcumin can quench singlet oxygen and can inhibit the production of H_2O_2 in an aqueous environment [143]. According to Balasubramanyam et al., the antioxidant and free radical scavenging properties of gold nanoparticles@curcumin over free curcumin may be due to their interaction patterns of phenolic OH and by abstraction of H-atom from C-12 methylene group of curcumin [144]. Further, the main reason for this phenomenon might be attributed to the increasing availability of curcumin conjugated nano-gold, which provides more H-atoms available to cells.

Pu et al. have encapsulated skillfully curcumin antioxidant using smart nanoparticles, and they managed the release of the nanoformulation by both oxidative stress and lowering pH level to the infected tissues to initiate inhibition of the lipopolysaccharide (LPS) stimulated macrophage's overproduction of ROS or RNS [145].

Xing et al. observed that curcumin nanoliposomes outperformed free curcumin in terms of stability against alkaline pH and metal ions, as well as storage stability at 4 °C. Curcumin nanoliposomes have demonstrated strong constant releasing properties and comparable cellular antioxidant activities, which are primarily owe to their lower absorption by cells [146]. Shah et al. also found higher efficiency of radical scavenging activity of curcumin encapsulated by chitosan-tripolyphosphate nanoemulsion in comparison to free curcumin, which illustrates the emulsion's protective impact on curcumin bioactive compounds' antioxidant [147].

5.3 Quercetin, Catechin, and Resveratrol

Polyphenolic compounds such as quercetin and catechin are being used in pharmaceutical and food industries to slow down the oxidation of edibles or the human body. Antioxidant, antiradical, anti-inflammatory, antimutagenic, anticarcinogenic, antiangiogenic, antibacterial, antiviral, antiaging effects, and activating antioxidant enzymes in the human body are some of their beneficial properties. Hector Pool et al. have demonstrated that quercetin and catechin have antiradical effects, as they can obstruct lipid oxidation reactions in various processes.

They have developed quercetin and catechin encapsulated poly (D, L-lactide-co-glycolide) (PLGA) nanoparticles and released quercetin and catechin in a regulated manner to exploit their antioxidant properties [148]. Wu et al. have reported that the antioxidant activity of eudragit E encapsulated quercetin nanoparticle along with its anti-superoxide formation, DPPH superoxide anion scavenging, and anti-lipid peroxidation activities are all more efficacious than found with bare free quercetin [149].

As a flavonoid, quercetin is found to contribute to the utmost antioxidant action through multiple pathways The use of quercetin as a direct topical treatment reduced the skin damage caused by UV exposure [150]. But because of its low permeation through skin, the compound's skin-protective function becomes limited by a lack of adequate quantity of the compound reaching the epidermis' real spot of action.

Quercetin has been formulated in a variety of ways, including cocooned in microemulsions and lipid nanoparticle and enveloping in silica nanoparticle, to improve skin penetration. It was found that the emulsion having silica nanoparticles served the only carrier capable of significantly increasing quercetin in vivo penetration into the human *stratum corneum* while keeping silica particles restricted only to the upper layers [151].

To boost oral biological availability and antioxidant capability, Jain et al. have developed a self-emulsifying drug delivery system (SEDDS). The antioxidant activity of quercetin-SEDDS was again found to be higher than that of free quercetin in a DPPH scavenging assay [152].

Carlotti et al. *have* developed resveratrol-loaded solid lipid nanoparticles for testing the anti-lipoperoxidative activity by using malondialdehyde and thiobarbituric. The protective effect of resveratrol against UV-radiation induced skin damage in porcine skin was observed 5.6-fold greater (increased antioxidant activity) with resveratrol-loaded solid lipid nanoparticles [153].

5.4 Zein

Zein is a prolamine-protein contained in maize that has long been used in the food and pharmaceutical industries due to its superior biodegradability and biocompatibility and ability to shape films. Under acidic conditions, zein tends to form a film, while under neutral and basic conditions, it tends to form particles. The ability of zein to encapsulate essential oils such as oregano, red thyme, and cassia is due to its unusual solubility in binary ethanol aqueous solutions. Recently, the antioxidant activity and the rate kinetics of release of essential oils encapsulated in zein nanoparticles were studied intensely [154].

5.5 Chitosan

Chitosan is a natural hetero-polymer derivative of chitin, which is commonly used to provide a nanomaterial a positive surface charge. It has been found to provide a stable resource for encapsulation. Acquiring positive zeta potential of the chitosan nanoparticle, cellular uptake of nanoparticles increases tremendously [155].

With the help of various proven in vitro methods, including metal ion's chelating and reducing capacity, the superoxide and hydroxyl radical scavenging assays, Natthan et al. investigated the reducing capabilities of various chitosan complex solutions with various ligands such as acetate (CS-acetate), hydroxybenzotriazole (CS-HOBt), thiamine pyrophosphate (CS-TPP), and ethylenediaminetetraacetic acid (CS-EDTA). Among these complexes of chitosan, the extreme superoxide radical scavenging effect was seen with CS-acetate concentration (IC50) equal 0.349 mg/mL, while the maximum hydroxyl radical preying effect was seen with CS-HOBt concentration (IC50) at 0.34 mg/mL. In terms of metal ion chelating activity, CS-EDTA had the maximum (approximately 100% at 1 mg/mL), while the others had just only 20% activity at the same concentration. The CS-TPP had shown the maximum reducing capacity, according to their findings [156].

Selenium nanoparticles stabilized with chitosan (CS-SeNPs) showed a lower toxicity and can penetrate the tissues to perform antioxidant effects that are more noticeabe in inernals than in skin. They found that a 30-day storage process could enhance the antioxidant capacities. The good abilities to penetrate cell or tissue have made these nanoparticles to be able to effectively inhibit ROS accumulation, reduce Se cytotoxicity, protect glutathione peroxidase activity, and prevent Lipofuscin accumulation, in vitro or in vivo [157].

The chitosan-templated PB NPs coordinated with the optimal chitosan molecular weight had uniform sphere-like particles, improved stability, and effective scavenging activity of in vitro reactive oxygen species generation in murine fibroblast cells stimulated by oxidative stress agents without any cytotoxicity, implying that they could be promising antioxidant agents [158].

Muco-adhesive property of chitosan have an additional advantage due to which it may be durably used with nanoparticles to be operated in mucosal surfaces in enhancing the targeted delivery of nanoparticles.

Nayak et al. exploited this property of chitosan in framing nanoformulations. They have synthesized silver nanoparticles through green routes encapsulating polyphenolic antioxidants viz. Vitamin C (ascorbic acid), Vitamin E (tocopherol), and benzene-1,2-diol (catechol). They followed ionotropic gelation method to enable encapsulation process and avoid instability that may occur due to redundant initial burst release and incomplete release. Additionally, it enables the antioxidant drugs to be delivered to breast cancer cells in a targeted manner to speed up the scavenging of reactive oxygen species (ROS) produced in large amount. These nanoformulations displayed a high degree of encapsulation (up to 76%) and antioxidant activity compared to bulk equivalents. Moreover, such synergism created minimal side effects to the normal cells as they only released the drug for a persisting long duration at the specific locations [159].

5.6 Gallic Acid

Gallic acid (GA) is a popular and well-known robust natural antioxidant found in a variety of herbs. Because of its unique physiochemical and economical properties, such as having innocuous nature, biologically degradability, ample availability, and a low price, it avails a wide range of applications in medicine, food, and pharmacological industries. Since GA is having multi-therapeutical approaches, it facilities the antioxidant, anti-inflammatory, anticancer, antitumor, antimicrobial, and antidiabetic properties. Surface-functionalized nanomaterials have demonstrated that attachment of the antioxidants results in increased antioxidant activity and bioavailability. The successful functionalization of GA on silica nanoparticles surface was identified as an efficient nano-antioxidant by authors. The bimetallic (Ag-Se) nanoparticles functionalized with quercetin and gallic acid were used as antioxidant, antimicrobial, and antitumor agents [40, 43].

6 Nanoemulsions and Nanocarriers

Emulsions can be prepared by dispersing two immiscible liquids, such as water and oil. Oil droplets get dispersed in aqueous phase, and water surrounding it forms the continuous phase, oil in water (O/W) or water in oil (W/O). Nanoemulsions are easily produced in large quantities by high shear stress or mechanical extrusion process and can be engineered with specific attributes such as size, surface charge, prolonged blood circulation and targeting properties. Nanoemulsions can be prepared in order to tune their physicochemical properties to be applied in therapy, diagnostics, or theranostics. The size of the droplets varies depending on the drug particles, mechanical energy, composition, and relative amount of the surfactants. Nanoemulsions are sometimes known as mini-emulsions, fine-dispersed emulsions, submicron emulsions, etc. based on the size of the droplets of dispersed phase. The amount of oil in O/W nanoemulsions may vary but generally lies within 5–20% w/w. Sometimes a mixture of oils is used to improve drug solubilization in the oil phase. A co-surfactant or a co-solvent may also be used in addition to the surfactant to facilitate the stabilization process [160, 161]. Nanoemulsions, by virtue of their lipophilic nature and low globule size, are widely explored as a delivery system to enhance uptake across the nasal mucosa. The possibility of using muco-adhesive agents such as polyelectrolyte polymers can prolong the nasal mucosa interaction, providing an extended delivery of the drug to the olfactory region and henceforth to the brain [162].

Natural nanoparticles like liposomes, niosomes, etc. and synthetic polymeric nanoparticles such as PEG-copolymers have emerged as efficient nanocarriers for antioxidant delivery and enhancement of antioxidant activity taking nano-antioxidants into the domain of new clinical applications with increased surface area, higher solubility, improved stability, controlled release of active ingredients, reducing skin irritancy, protection from degradation and aging, increased drug (antioxidants) loading, and improved permeation of actives components into the various target cells [163].

7 Determination of Antioxidant Activity

Antioxidant activity could be determined simply by placing the fat in a closed container with oxygen and measuring the rate of oxygen consumption. However, the possible mechanisms of action of antioxidants were first explored when it was recognized that a substance with anti-oxidative activity is likely to be one that is itself readily oxidized. Researches showed that the antioxidants prevent the process of lipid peroxidation, which led to the identification of antioxidants as reducing agents that prevent oxidative reactions by neutralizing free radicals and scavenging reactive oxygen species before they can damage cells.

Excess free radicals can lead to a condition called oxidative stress. Oxidative stress means an imbalance between pro-oxidants and antioxidant mechanisms, which causes due to excessive oxidative metabolism. When the body does not have enough natural antioxidants to protect it from the reactive oxygen based free radicals that we produce, our cell tissues can be attacked by the excess free radicals present, resulting in oxidative stress. This stress can be resulted due to several environmental factors mentioned earlier such as exposure to pollutants, alcohol, medications, infections, poor diet, toxins, radiation, etc. Oxidative damage to DNA, proteins, and other macromolecules may lead to a wide range of human diseases most notably heart disease and cancer.

Oxidative stress measurement provides critical measures of how effectively antioxidants are protecting cells from excess free radicals. It is also a measure of how our antioxidants are protecting us from excess free radicals. There are number of assays through which antioxidant activity of metals or compounds can be determined. These assays are attractive for their simplicity, convenience, reproducibility, and low cost.

7.1 Parameters to Measure Antioxidant Ability

Generally in-vitro evaluation techniques involve spectrophotometry, fluorescence, chemiluminescence, chromatography, electron spin resonance, etc. Some of the parameters that is determined to measure the antioxidant ability of a particular analyte are mentioned below:

1. Total antioxidant activity.
2. Total antioxidant capacity.
3. Total antioxidant potentials.
4. Trolox equivalent antioxidant capacity.
5. Total radical absorption potentials.
6. Ferric reducing/antioxidant power.
7. Oxygen radical absorption capacity.

All these measurements are based on either a hydrogen atom transfer (HAT) reaction or a single electron transfer (SET) reaction from an antioxidant or oxidant to a free radical [164–176].

The various parameters that determined under HAT and SET methods are mentioned under following heads accordingly:

7.2 Hydrogen Atom Transfer (HAT) Methods

1. Oxygen radical absorbance capacity (ORAC).
2. Lipid peroxidation inhibition capacity (LPIC).
3. Total radical trapping antioxidant parameter (TRAP).
4. Inhibited oxygen uptake (IOC).
5. Crocin bleaching nitric oxide radical inhibition activity.
6. Hydroxyl radical scavenging activity by p-NDA(p-butrisidunethyl aniline).
7. Scavenging of H_2O_2 radicals.
8. ABTS radical scavenging.
9. Scavenging of superoxide radical formation by alkaline (SASA).

7.3 *Single Electron Transfer (SET) Methods*

1. Trolox equivalent antioxidant capacity (TEAC) decolorization.
2. Ferric reducing antioxidant power (FRAP).
3. DPPH free radical scavenging.
4. Copper(II) reduction capacity.
5. Total phenols by Folin-Ciocalteu.
6. N,N-dimethyl-p-phenylenediamine (DMPD).

The change of optical absorbance of either antioxidant or oxidant is measured for the quantitation for antioxidant capability. The HAT-based assays measure the capability of an antioxidant to quench free radicals (generally, biologically more relevant peroxyl radicals are considered) by H-atom donation. While in the SET-based assays, the antioxidant action is simulated with a suitable redox-potential probe, namely, the antioxidants react with a fluorescent or colored probe (oxidizing agent) instead of peroxyl radicals. This spectrophotometric SET-based assays measure the capacity of an antioxidant in the reduction of an oxidant, which changes color when reduced. The degree of color change (either an increase or decrease of absorbance of the probe at a given wavelength) is correlated to the concentration of antioxidants in the sample analytes say in ABTS or Trolox, etc. The chemical approaches are simple and easy to study the total antioxidant activity, which includes the radical scavenging ability and reductive activity. Radicals are commonly used to measure the radical scavenging ability of antioxidant.

Now the antioxidant capacity assays to measure antioxidant capacity are categorized according to the mode of measurement [177]:

7.3.1 Spectrometric Assays

Spectrophotometric assays measure the capacity of antioxidant nanoparticles in the reduction of an oxidant, which changes color when reduced. The degree of color change (either an increase or decrease of absorbance at a given wavelength) can be correlated with the concentration of antioxidants in the sample. These can be categorized as:

1. DPPH Assay: In this assay, the color change of antioxidant reaction with an organic radical 2,2-di (4-tert-octylphenyl)-1-picrylhydrazyl(DPPH), rather decolorization is observed colorimetrically.
2. ABTS Assay: The color change of antioxidant reaction with an organic cation radical, 2,2′-Azinobis-(3-ethylbenzothiazoline-6-sulfonicacid, (ABTS) and decolorization is observed colorimetrically.
3. FRAP Assay: The antioxidant reaction with a Fe(III) complex, ferric reducing antioxidant power (FRAP) rather increase in absorbance at particular wave length is observed colorimetrically.

4. PFRAP Assay: The reduction of potassium ferricyanide by antioxidants and subsequent reaction of potassium ferrocyanide with Fe^{3+} is observed colorimetrically.
5. CUPRAC Assay: Here reduction of Cu (II) to Cu (I) occurs by antioxidants and color change is detected colorimetrically. Increase in the color intensity occurs due to the formation of charge transfer spectra by Cu(I).
6. ORAC Assay: The antioxidant capacity is determined by adding sample to the peroxyl radical generator, 2,2′-azobis(2-amidinopropane)dihydrochloride (AAPH). The inhibition of the free radical action is measured using the fluorescent compound, B-phycoerythrin or R-phycoerythrin, with peroxyl radicals is observed by the loss of fluorescence of fluorescein which is induced by AAPH.
7. HORAC Assay: The antioxidant capacity is observed as the capacity to quench OH radicals generated by a Co(II) based Fenton-like system. The effect is assayed by the loss of fluorescence of fluorescein.
8. TRAP Assay: Here the antioxidant capacity to scavenge luminol-derived radicals is assayed by chemiluminescence quenching which is generated from AAPH decomposition.
9. Fluorimetric Assay: Emission of light by a substance that has absorbed light or other electromagnetic radiation of a different wavelength is done by recording of fluorescence excitation/emission spectra.

7.3.2 Electrochemical Techniques

1. By Cyclic voltammetry: Here, the potential of a working electrode is linearly varied from an initial value to a final value and back, and the respective current intensity is recorded by the measurement of the intensity of the cathodic/anodic peak.
2. By Amperometry: In this technique, the potential of the working electrode is set at a fixed value with respect to a reference electrode and the measurement of the intensity of the current generated by the oxidation/reduction of an electroactive analyte is done.
3. By Biamperometry: Here the reaction of the analyte (antioxidant) with the oxidized form of a reversible indicating redox couple is studied. The current flowing between two identical working electrodes is measured at a small potential difference that are immersed in a solution containing the analysed sample and a reversible redox couple.

7.3.3 Chromatographic Techniques

1. Gas chromatography: In this technique, the separation of the compounds in a mixture is based on the repartition between a liquid stationary phase and a gas mobile phase. The compounds which are analyzed for the antioxidant activity are lipid peroxides, aldehydes, tocopherols, sterols, phenolic acids, flavonoids, etc. With the help of flame ionization or through thermal conductivity the end product is detected.

2. High performance liquid chromatography: In this technique, the separation of the compounds in a mixture is based on the repartition between a solid stationary phase and a liquid mobile phase with different polarities at high flow rate and pressure of the mobile phase. Through this technique, antioxidants analysis of flavonoid, tocopherols, aldehydes, phenolic acids, etc. can be done. The final end product is detected through UV–Vis (e.g., diode array), fluorescence, mass spectrometry or electrochemically.

Among all the abovementioned assays, DPPH method is the most popular one. Descriptions of few other methods are also given in subsequent paragraphs.

7.3.4 DPPH Method

DPPH is intensely colored crystalline substance (λmax 520 nm), stable in alcohol solutions. When storing it in ethanol solution in the dark for 240 hours, the intensity of the maximum absorption of its electronic spectrum remains unchanged. It is important that the DPPH is stable to air oxygen [178]. These properties are important when choosing a method of determining the antioxidant activity of medicinal compounds for their reactivity with DPPH. According to Trogadas et al., an appropriate amount of solution (stock solution in toluene) of the metal nanoparticles was dispersed in 50 mL ethanol to obtain a suspension with a concentration of 10 mg/L. Then, 1 mL of metal suspension (Au, Pd, Pt or Ag) was added to 2 mL DPPH solution in a test tube. The test tube was heated to 80 °C in a thermostatic bath to increase the rate of reaction; the reaction was quenched after 30 min by cooling down the mixture in an ice bath. When the DPPH radical is scavenged by an antioxidant (metal nanoparticles), it transforms to 1,1-diphenyl-2-picrylhydrazyne (DPPH-H). As part of this transformation, the color of the solution turns from purple to yellow (Fig. 3.3). The extent of the transformation (i.e., radical scavenging) was quantified by the decay in absorbance at 523 nm over the duration of the test [179] (Fig. 3.12).

Trogadas et al. found that Pt nanoparticles demonstrated the highest activity against DPPH radicals as a 50% decrease in UV–Vis absorbance was achieved and almost double the rate of reaction was obtained when compared with Pd and Au nanoparticles. The scavenging activity of metal nanoparticles against the DPPH free radical was also established by them using electronic paramagnetic resonance (EPR) also known as electron spin resonance (ESR). The samples were prepared and treated as described above but the DPPH free radical was detected using electron spin resonance (ESR) spectroscopy. When the DPPH radical is scavenged by an antioxidant, it transforms to 1,1-diphenyl-2-picrylhydrazyne (DPPH-H). The extent of the transformation (i.e., radical scavenging) was quantified by the decay in ESR signal.

$$\text{Radical Scavenging Effect } (\%) = \left(1 - \text{A}_{\text{samples,523nm}}/\text{A}_{\text{control,523nm}}\right) \times 100$$

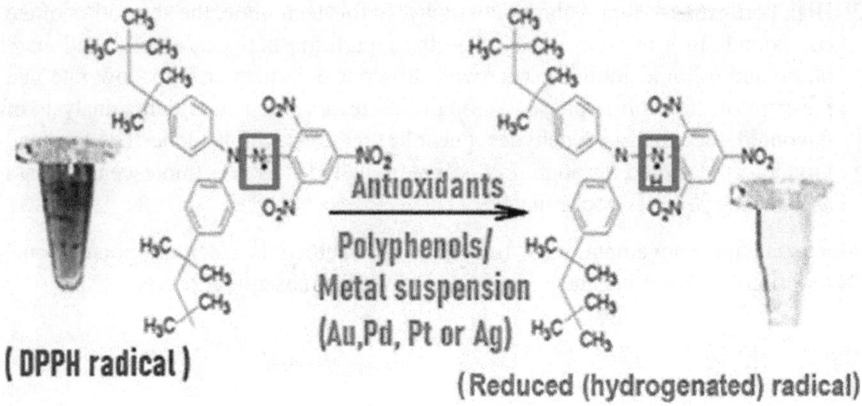

Fig. 3.12 Representation of DPPH radical activity

7.3.5 Aldehydic Secondary by-Products of Lipid Peroxidation as Markers of Oxidative Stress

Lipid peroxidation is a well-defined mechanism of cellular damage in animals and plants. Lipid peroxides are unstable indicators of oxidative stress in cells that decompose to form more complex and reactive compounds such as Malondialdehyde (MDA), a three-carbon dialdehyde and 4-hydroxynonenal (4-HNE), natural by-products of polyunsaturated lipid peroxidation. Aldehydes are the by-products of oxidative stress. Measuring the end products of lipid peroxidation is one of the most widely accepted assays for oxidative damage. These aldehydic secondary by-products of lipid peroxidation are generally accepted as markers of oxidative stress, i.e., it is an effective biomarker of excess free radicals.

Generally aldehyde measurement is done in exhaled breath condensate and urine. Subsequent studies have shown a correlation between breath aldehydes and disease states. MDA is a low-molecular-weight end products formed via the decomposition of certain primary and secondary lipid peroxidation products. It can be found in most biological samples including foodstuffs, serum, plasma, tissues, and urine, as a result of lipid peroxidation. It is a highly reactive compound that is not typically observed in pure form. In organic solvents, the *cis*-isomer is favored, whereas in water the *trans*-isomer predominates. In the laboratory it can be generated in situ by hydrolysis of 1,1,3,3-tetramethoxypropane, which is commercially available. It mainly exists in the enol form:

$$CH_2(CHO)_2 \rightarrow HOCH = CH - CHO$$

MDA is a reactive oxygen species (ROS), and as such is assayed in vivo as a biomarker of oxidative stress. MDA and other thiobarbituric reactive substances

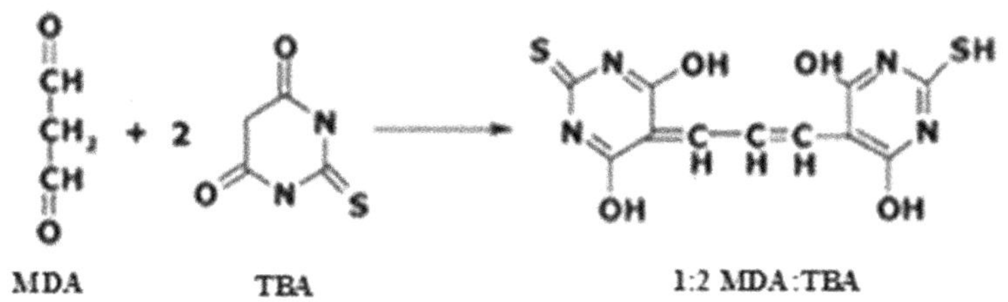

Scheme 3.1 Formation of 1:2 MDA:TBA complex

(TBARS) condense with two equivalents of thiobarbituric acid to give a fluorescent red derivative, that can be assayed spectrophotometrically. At low pH and elevated temperature, MDA readily participates in nucleophilic addition reaction with 2-thiobarbituric acid (TBA), generating a red, fluorescent 1:2 MDA:TBA complex adduct which absorbs strongly at 532 nm [180] (Scheme 3.1).

7.3.6 Chemiluminescence Assay

The in vitro antioxidant activity of plant extracts and of herbal silver nanoparticles has been determined by chemiluminescence (CL) measurements using a chemiluminometer by Lacatusu et al. [181]. Luminol—a cyclic hydrazide, has been used as a light amplifying substance which emits light when oxidized in the presence of oxidizing species. As a free radicals generator system, it has been used H_2O_2 in TRIS-HCl solution buffer (pH = 8.6). The antioxidant activity (percentage of free radical scavenging) of each sample was calculated using the expression:

$$AA = I_0 - I/I_0 \times 100\%$$

where I_0 is the maximum CL intensity for *standard* (reaction mixture without sample) at t = 5 seconds and I is the maximum CL intensity for sample at t = 5 seconds.

7.3.7 β-Carotene Bleaching Assay

This method was developed by Wettasinghe et al. [182]. For determining antioxidant activity by this method, 2 ml of β-carotene solution (0.2 mg/ml in chloroform) is generally pipetted out into a round bottom flask containing 20 linoleic acid and 200μl nonionic detergents. The mixture is then evaporated at 40 °C for 10 min to remove the solvent, immediately followed by the addition of distilled water (100 ml). After agitating vigorously the mixture, 5 ml aliquots of the resulting emulsion has to be transferred into test tubes containing different concentrations (5–20 mg/ml) of extracts. The mixture will be vortexed and placed in a water bath at 50 °C for 2 h, while the absorbance of the tested sample will repeatedly be measured by every 15 min at 470 nm using a UV–Vis spectrophotometer. For blank, the solution should contain the same concentration of sample without β-carotene.

All determinations have to be performed in triplicates and then total antioxidant activity will be calculated based on the following equation:

$$AA = 1 - (A_0 - A_t)/\left(A_0{}^0 - A_0{}^t\right)$$

where AA is the antioxidant activity, A_0 and $A_0{}^0$ are the absorbance values measured at initial time of the incubation for samples and control, respectively, while A_t and $A_0{}^t$ are the absorbance in the samples and control at $t = 120$ min.

7.3.8 Oxygen-Radical Absorbance Capacity (ORAC) Assay

With the help of electron spin resonance (ESR), spectroscopic technique, one can directly detect free radical species in the sample. However, several other methods are also there based on elimination of oxygen radicals by antioxidants. The oxygen-radical absorbance capacity (ORAC) assay is based on the ability of a sample to slow down or terminate the radical reaction. The assay is based on the decrease in the fluorescence of protein β-phycoerythrin upon peroxyl radical attack. These radicals are generated by AAPH [2,2′-azobis(2-amidinopropane) dihydrochloride]. The decrease in fluorescence of β-phycoerythrin in the presence of AAPH shows linear relationship with time and a period of complete protection of β-phycoerythrin against AAPH is correlated to antioxidant concentrations [183]. The obtained results are compared with the activity of standard Trolox and expressed as ORAC units. 1 ORAC unit is equal to the net protection produced by 1μM Trolox. In this assay, the total antioxidant capacity of a sample is estimated by taking the oxidation reaction to completion [184].

7.3.9 Ferric Reducing Antioxidant Power (FRAP) Assay

Ferric reducing antioxidant power (FRAP) assay is one of the important method to determine antioxidant capacity chemically. It measures the change in absorbance at 593 nm, from colorless oxidized Fe(III)form to the formation of a blue colored Fe (II)-tripyridyltriazine compound, by the act of electron donating antioxidants [185]. This assay has several limitations (as non-physiological pH) and reflects only the ability of tested compound to reduce Fe(III) ions.

8 Cytotoxicity

Some nanoparticles have been shown to stimulate ROS production and cause oxidative damage in animals. Some studies in rodents have shown that single-wall carbon nanotubes are cytotoxic [186, 187]. Soto et al. examined the cytotoxicity of a well-characterized group of nanoparticles on an established cell line of murine macrophages and showed that a variety of nanoparticles are cytotoxic [188]. In addition, it has been shown that water-soluble fullerenes are able to directly generate superoxide anions that are also cytotoxic [189]. Though studies have been shown that fullerene derivatives act as strong antioxidants in solutions, but the data obtained on cell cultures etc. are controversial concerning their antioxidant properties. Certain studies confirm the antioxidative action of fullerenes, whereas others show that cells treated with fullerenes exhibit signs of oxidative stressin animals [190].

9 Conclusion

Nanomaterials no doubt are charismatic. Their increased surface to volume ratio makes them very precious in terms of exploitation of their physicochemical properties for various applications. Nano-metals and nanomaterials can act as excellent antioxidants by scavenging free electrons of reactive organic species and free radicals produced in our body as they have capacity to decompose these free radicals and ROS. Scientists have suggested several mechanisms for the antioxidant activity of these moieties. But a basic of all the mechanisms is the promulgation of redox reactions between ROS and antioxidants. Out of the number of determination methods for antioxidant activity, three methods are commonly used in practice, which were described briefly. A review by Gonzalez et al. reports the use of electron spin resonance to assess the antioxidant ability by different methods in biological systems in detail [191].

Certain nanomaterials have also proven to be toxic by stimulating ROS production and causing oxidative damages. Water-soluble fullerenes have also shown toxicity to certain extent. A meticulous study is still required to avail the wholesome benefits of nano-metals and nanomaterials. But there is no doubt on the characteristic feature of antioxidant that has been recognized as valuable additives for averting acute and chronic diseases.

Jennifer E. Slemmer and John T. Weber, Limitations in Neuroprotective Strategies, *Antioxidants* 2014, *3*(4), 636–648.

References

1. Sanjay SS, Pandey AC. In: Shukla AK, editor. A brief manifestation of nanotechnology, in EMR/ESR/EPR spectroscopy for characterization of nanomaterials, advanced structured materials. © Springer (India) Pvt. Ltd.; 2017. p. 47–64. https://doi.org/10.1007/978-81-322-3655-9_2.
2. Li Y, Lin R, Wang L, Huang J, Wu H, Cheng G, Zhou Z, MacDonald T, Yang L, Mao H. Peg-bage polymer coated magnetic nanoparticle probes with facile functionalization and anti-fouling properties for reducing non-specific uptake and improving biomarker targeting. J Mater Chem B. 2015;3:3591–603.
3. Santiago-Rodriguez L, Lafontaine MM, Castro C, Mendez-Vega J, Latorre-Esteves M, Juan EJ, Mora E, Torres-Lugo M, Rinaldi C. Synthesis, stability, cellular uptake, and blood circulation time of carboxymethyl-inulin coated magnetic nanoparticles. J Mater Chem B. 2013;1:2807–17.
4. Majeed MI, Lu Q, Yan W, Li Z, Hussain I, Tahir MN, Tremel W, Tan B. Highly water-soluble magnetic iron oxide (fe3o4) nanoparticles for drug delivery: enhanced in vitro therapeutic efficacy of doxorubicin and mion conjugates. J Mater Chem B. 2013;1:2874–84.
5. Nasirimoghaddam S, Zeinali S, Sabbaghi S. Chitosan coated magnetic nanoparticles as nano-adsorbent for efficient removal of mercury contents from industrial aqueous and oily samples. J Ind Eng Chem. 2015;27:79–87.
6. Deligiannakis Y, Sotiriou GA, Pratsinis SE. Antioxidant and antiradical sio_2 nanoparticles covalently functionalized with gallic acid. ACS Appl Mater Interfaces. 2012;4:6609–17.
7. Bhattacharya K, Gogoi B, Buragohain AK, Deb P. Fe2o3/c nanocomposites having distinctive antioxidant activity and hemolysis prevention efficiency. Mater Sci Eng C. 2014;42:595–600.
8. Sodipo BK, Abdul Aziz A. Non-seeded synthesis and characterization of superparamagnetic iron oxide nanoparticles incorporated into silica nanoparticles via ultrasound. Ultrason Sonochem. 2015;23:354–9.
9. Gupta AK, Gupta M. Synthesis and surface engineering of iron oxide nanoparticles for biomedical applications. Biomaterials. 2005;26:3995–4021.
10. Chorny M, Hood E, Levy RJ, Muzykantov VR. Endothelial delivery of antioxidant enzymes loaded into non-polymeric magnetic nanoparticles. J Control Release. 2010;146:144–51.
11. Lacramioara L, Diaconu A, Butnaru M, Verestiuc L. Biocompatible spions with superoxide dismutase/catalase immobilized for cardiovascular applications. In: Sontea V, Tiginyanu I, editors. Third International Conference on Nanotechnologies and Biomedical Engineering, vol. 55. Singapore: Springer; 2016. p. 323–6.
12. Laurent S, Forge D, Port M, Roch A, Robic C, Vander Elst L, Muller RN. Magnetic iron oxide nanoparticles: synthesis, stabilization, vectorization, physicochemical characterizations, and biological applications. Chem Rev. 2008;108:2064–110.
13. Abdel-Aziz MS, Shaheen MS, El-Nekeety AA, Abdel-Wahhab MA. J Saudi Chem Soc. 2014;18:356.
14. Martın R, Menchon C, Apostolova N, Victor VM, Alvaro M, Herance JR, Garcia H. ACS Nano. 2010;4:6957.
15. Lordan S, Mackrill JJ, et al. J Nutri Biochem. 2009;20(5):321–36.
16. Halliwell B, Gutteridge JMC. Free radicals, other reactive species and disease. In: Free radicals in biology and medicine. New York: Oxford Univ. Press; 1999. p. 617–783.
17. Sies H. Antioxidants in disease mechanisms and therapy. San Diego: Academic Press; 1997.
18. Brenneisen P, Steinbrenner H, Sies H. Mol Asp Med. 2005;26:256.
19. Cornelli U. Clin Dermatol. 2009;27:175.
20. Margaill M, Plotkine DL. Free Radic Biol Med. 2005;39:429–43.
21. Pryor WA. Free Radic Biol Med. 2000;28:141–64.
22. Kaliora C, Dedoussis GVZ, Schmidt H. Atherosclerosis. 2006;187:1–17.
23. Shahidi F, Ho CT. Antioxidant measurement and applications. ACS Symp Ser. 2007;956:2–7. Chapter 1

24. Wolfe KL, Liu RH, Agric J. Food Chem. 2007;55:8896–907.
25. Ma XY, Li H, Dong J, Qian WP. Food Chem. 2011;126:698–704.
26. Awika JM, Rooney LW, Wu X, Prior RL, Zevallos LC. J Agric Food Chem. 2003;51:6657–62.
27. Korkina LG, Afans'ev IB. Antioxidant and chelating properties of flavonoids. Adv Pharmacol. 1997;38:151–63.
28. Slemmer JE, Weber JT. Limitations in neuroprotective strategies. Antioxidants. 2014;3 (4):636–48.
29. Sharpe E, Andreescu D, Andreescu S. Artificial nanoparticle antioxidants. In oxidative stress: diagnostics, prevention, and therapy. Washington, DC: ACS Publications; 2011. p. 235–53.
30. Dikpati A, Madgulkar AR, Kshirsagar SJ, Bhalekar MR, Singh CA. Targeted drug delivery to CNS using nanoparticles. J Adv Pharm Sci. 2012;2:179–91.
31. Yin HY. Free Radic Biol Med. 2007;43:1229–30.
32. Seeram NP, Aviram M, Zhang YJ, Henning SM, Feng L, Dreher M, Heber D. J Agric Food Chem. 2008;56:1415–22.
33. Moore J, Yin JJ, Yu LL. J Agric Food Chem. 2006;54:617–26.
34. Paciotti GF, Myer L, Weinreich D, Goia D, Pavel N, McLaughlin RE, Tamarkin L. Drug Deliv. 2004;11:169–83.
35. Bielinska JD, Eichman I, Lee JR, Baker LB. J Nanopart Res. 2002;4:395–403.
36. Roth J. Cell Biol. 1996;106:79–92.
37. Li H, Ma XY, Dong J, Qian WP. Anal Chem. 2009;81:8916–22.
38. Leonard K, Ahmmad B, Okamura H, Kurawaki J. Colloids Surf B Biointerfaces. 2011;82:391–6.
39. Ma XY, Qian WP. Biosens Bioelectron. 2010;26:1049–55.
40. Scampicchio M, Wang J, Blasco AJ, Arribas AS, Mannino S, Escarpa A. Anal Chem. 2006;78:2060–3.
41. Mannino S, Brenna O, Buratti S, Cosio MS. Electroanalysis. 1998;10:908.
42. Blasco AJ, Rogerio MC, Gonzalez MC, Escarpa A. Anal Chim Acta. 2005;539:237.
43. Xia D-Z, Yu X-F, Zhu Z-Y, Zou Z-D. Nat Prod Res. 2011;25:1893–901.
44. Nie Z, Liu KJ, Zhong CJ, Wang LF, Yang Y, Tian Q, Liu Y. Enhanced radical scavenging activity by antioxidant-functionalized gold nanoparticles: a novel inspiration for development of new artificial antioxidants. Free Radic Biol Med. 2007;43:1243–54.
45. Libo D, Suo S, Wang G, Jia H, Liu KJ, Zhao B, Liu Y. Mechanism and cellular kinetic studies of the enhancement of antioxidant activity by using surface-functionalized gold nanoparticles. Chem Eur J. 2013;19:1281–7. https://doi.org/10.1002/chem.201203506.
46. Vilas V, Philip D, Mathew J. Essential oil mediated synthesis of silver nanocrystals for environmental, anti-microbial and antioxidant applications. Mater Sci Eng C. 2016;61:429–36.
47. Marulasiddeshwara M, Dakshayani S, Kumar MS, Chethana R, Kumar PR, Devaraja S. Facile-one pot-green synthesis, antibacterial, antifungal, antioxidant and antiplatelet activities of lignin capped silver nanoparticles: a promising therapeutic agent. Mater Sci Eng C. 2017;81:182–90.
48. Sriranjani R, Srinithya B, Vellingiri V, Brindha P, Anthony SP, Sivasubramanian A, Muthuraman M. S silver nanoparticle synthesis using Clerodendrum phlomidis leaf extract and preliminary investigation of its antioxidant and anticancer activities. J Mol Liq. 2016;220:926–30.
49. Kalaiyarasan T, Bharti VK, Chaurasia O. One pot green preparation of Seabuckthorn silver nanoparticles (SBT@ AgNPs) featuring high stability and longevity, antibacterial, antioxidant potential: a nano disinfectant future perspective. RSC Adv. 2017;7:51130–41.
50. Teerasong S, Jinnarak A, Chaneam S, Wilairat P, Nacapricha D. Poly (vinyl alcohol) capped silver nanoparticles for antioxidant assay based on seed-mediated nanoparticle growth. Talanta. 2017;170:193–8.
51. Gao X, Zhang J, Zhang L. Adv Mater. 2002;14:290.

52. Mishra M, Kumar H, Tripathi K. Diabetic delayed wound healing and the role of silver nanoparticles. Dig J Nanomater Biostruct. 2008;3(2):49.
53. Mittal AK, Kaler A, Banerjee UC. Free radical scavenging and antioxidant activity of silver nanoparticles synthesized from flower extract of *Rhododendron dauricum*. Nano Biomed Eng. 2012;4(3):118–24. https://doi.org/10.5101/nbe.v4i3.p118-124.
54. Mittal AK, Kumar S, Banerjee UC. Quercetin and gallic acid mediated synthesis of bimetallic (silver and selenium) nanoparticles and their antitumor and antimicrobial potential. J Coll Interface Sci. 2014;431:194–9.
55. Sotiriou GA, Blattmann CO, Deligiannakis Y. Nanoantioxidant-driven plasmon enhanced proton-coupled electron transfer. Nanoscale. 2016;8:796–803.
56. Estevez AY, Erlichman JS. Cerium oxide nanoparticles for the treatment of neurological oxidative stress diseases. In: Andreescu ES, Hempel M, editors. Oxidative stress: diagnostics, prevention and therapy (volume 1083). Washington, DC: American Chemical Society; 2011. p. 255–88.
57. Estevez AY, Pritchard S, Harper K, et al. Neuroprotective mechanisms of cerium oxide nanoparticles in a mouse hippocampal brain slice model of ischemia. Free Radic Biol Med. 2011;51(6):1155–63.
58. Hirst SM, Karakoti AS, Tyler RD, Sriranganathan N, Seal S, Reilly CM. Anti-inflammatory properties of cerium oxide nanoparticles. Small. 2009;5(24):2848–56.
59. Pirmohamed T, Dowding JM, Singh S, et al. Nanoceria exhibit redox state-dependent catalase mimetic activity. Chem Commun (Camb). 2010;46(16):2736–8.
60. Korsvik C, Patil S, Seal S, Self WT. Superoxide dismutase mimetic properties exhibited by vacancy engineered ceria nanoparticles. Chem Commun (Camb). 2007;10(10):1056–8.
61. Das M, Patil S, Bhargava N, Kang JF, Riedel LM, Seal S, Hickman JJ. Biomaterials. 2007;28:1918–25.
62. Yu JC, Zhang L, Lin J. J Colloid Interface Sci. 2003;260:240–3.
63. Rzigalinski BA, Meehan K, Davis RM, Xu Y, Miles WC, Cohen CA. Nanomedicine. 2006;1:399–412.
64. Chen J, Patil S, Seal S, McGinnis JF. Nat Nanotechnol. 2006;1:142–50.
65. Niu J, Azfer A, Rogers LM, Wang X, Kolattukudy PE. Cardiovasc Res. 2007;73:549–59.
66. Colon J, Hsieh N, Ferguson A, Kupelian P, Seal S, Jenkins DW, Baker CH. Nanomedicine. 2010;6:698–705.
67. Kilbourn BT. In: Bloor D, Brook RJ, Flemings MC, Mahajan S, editors. Yttrium oxide, vol. 4. Oxford: Pergamon Press, Ltd.; 1994. p. 2957.
68. Kofstad P. Nonstoichiometry, diffusion, and electrical conductivity in binary metal oxides. New York: Wiley Interscience; 1972.
69. Kilbourn BT, Yttria. In: Bever MB, editor. Encyclopedia of materials science and engineering, vol. 7. Oxford: Pergamon Press, Ltd.; 1986. p. 5509.
70. Atou T, Kusaba K, Fukuoka K, Kikuchi M, Syono Y. Shockinduced phase transition of M_20_3 (M = Sc, Y, Sm, Gd, and In)-type compounds. J Solid State Chem. 1990;89:378.
71. Perebeinos V, Chan S-W, Zhang F. Solid State Commun. 2002;123:295.
72. Kingery WD, Bowen HK, Uhlman DR. Introduction to ceramics. second ed. New York: John Wiley; 1976.
73. Zhang F, Wang P, Koberstein J, Khalid S, Chan S-W. Surf Sci. 2004;563:74–82.
74. Kwon HJ, Cha M-Y, Kim D, Kim DK, Soh M, Shin K, Hyeon T, Mook-Jung I. Mitochondria-targeting ceria nanoparticles as antioxidants for Alzheimer's disease. ACS Nano. 2016;10 (2):2860–70. https://doi.org/10.1021/acsnano.5b08045.
75. Hochella MF Jr, Lower SK, Maurice PA, Pen RL, Sahai N, Sparks DL, Twining BS. Science. 2008;319:1631–5.
76. Radziun E, Dudkiewicz-Wilczyńska J, Ząbkowski T, et al. Assessment of the cytotoxicity of aluminum oxide nanoparticles on selected mammalian cells. Toxicol In Vitro. 2011;25 (8):1694–700.

77. Becker S, Soukup JM, Gallagher JE. Differential particulate air pollution induced oxidant stress in human granulocytes, monocytes and alveolar macrophages. Toxicol In Vitro. 2002;16:209.
78. Schubert D, Dargusch R, Raitano J, Chan S-W. Biochem Biophys Res Comm. 2006;342:86–91.
79. Tapeinos C, Battaglini M, Prato M, La Rosa G, Scarpellini A, Ciofani G. CeO_2 nanoparticles-loaded pH-responsive microparticles with antitumoral properties as therapeutic modulators for osteosarcoma. ACS Omega. 2018;3(8):8952–62.
80. Lee SS, Song W, Cho M, Puppala HL, Nguyen P, Zhu H, Segatori L, Colvin VL. Antioxidant properties of cerium oxide nanocrystals as a function of nanocrystal diameter and surface coating. ACS Nano. 2013;7(11):9693–703. https://doi.org/10.1021/nn4026806.
81. "Rice scientists create a super antioxidant"; news.rice.edu. 2018, 12–18.
82. Subbaiya R, Selvam M. Green synthesis of copper nanoparticles from Hibicus rosa-sinensis and their antimicrobial, antioxidant activities. Res J Pharm Biol Chem Sci. 2015;6:1183–90.
83. Din MI, Rehan R. Synthesis, characterization, and applications of copper nanoparticles. Anal Lett. 2017;50:50–62.
84. Das D, Nath BC, Phukon P, Dolui SK. Colloids Surf B Biointerfaces. 2013;101:430.
85. Ghosh S, More P, Nitnavare R, Jagtap S, Chippalkatti R, et al. Antidiabetic and antioxidant properties of copper nanoparticles synthesized by medicinal plant *Dioscorea bulbifera*. J Nanomed Nanotechnol. 2015;S6:007. https://doi.org/10.4172/2157-7439.S6-007.
86. Rawel HM, Frey SK, Meidtner K, Kroll J, Schweigert FJ. Determining the binding affinities of phenolic compounds to proteins by quenching of the intrinsic tryptophan fluorescence. Mol Nutr Food Res. 2006;50:705–13.
87. Ramasubbu N, Ragunath C, Mishra PJ, Thomas LM, Gyémánt G, et al. Human salivary alpha-amylase Trp58 situated at subsite −2 is critical for enzyme activity. Eur J Biochem. 2004;271:2517–29.
88. Yoshikawa Y, Hirata R, Yasui H, Sakurai H. Alpha-glucosidase inhibitory effect of anti-diabetic metal ions and their complexes. Biochimie. 2009;91:1339–41.
89. Wang Y, Xiang L, Wang C, Tang C, He X. Antidiabetic and antioxidant effects and phytochemicals of mulberry fruit (Morus alba L.) polyphenol enhanced extract. PLoS One. 2013;8:e71144.
90. Tripathi YB, Singh VP. Role of Tamra bhasma, an Ayurvedic preparation, in the management of lipid peroxidation in liver of albino rats. Indian J Exp Biol. 1996;34:66–70.
91. Sanyal AK, Pandey BL, Goel RK. The effect of a traditional preparation of copper, Tamra bhasma, on experimental ulcers and gastric secretion. J Ethnopharmacol. 1982;5:79–89.
92. Rakhmetova AA, Alekseeva TP, Bogoslovskaya OA, Leipunskii IO, Ol'khovskaya IP, et al. Wound-healing properties of copper nanoparticles as a function of physicochemical parameters. Nanotechnol Russ. 2010;5:271–6.
93. McMurry JF Jr. Wound healing with diabetes mellitus. Better glucose control for better wound healing in diabetes. Surg Clin North Am. 1984;64:769–78.
94. Rafique S, Idrees M, Nasim A, Akbar H, Athar A. Transition metal complexes as potential therapeutic agents. Biotechnol Mol Biol Rev. 2010;5:38–45.
95. Hatami M, Ghorbanpour M, Salehiarjomand H. Nano-anatase TiO2 modulates the germination behavior and seedling vigority of the five commercially important medicinal and aromatic plants. J Biol Environ Sci. 2014;8(22):53–9.
96. Ghorbanpour M, Hatam M, Hatami M. Activating antioxidant enzymes, hyoscyamine and scopolamine biosynthesis of *Hyoscyamus niger* L. plants with nano-sized titanium dioxide and bulk application. Acta agriculturae Slovenica. 2015;105(1) https://doi.org/10.14720/aas.2015.105.1.03.
97. Pedone D, Moglianetti M, De Luca E, Bardi G, Pompa PP. Platinum nanoparticles in nanobiomedicine. Chem Soc Rev. 2017;46:4951–75.

98. Liu Y, Wu H, Chong Y, Wamer WG, Xia Q, Cai L, Nie Z, Fu PP, Yin J-J. Platinum nanoparticles: efficient and stable catechol oxidase mimetics. ACS Appl Mater Interfaces. 2015;7(35):19709–17.
99. Chen C, Fan S, Li C, Chong Y, Tian X, Zheng J, Fu PP, Jiang X, Wamer WG, Yin J. Platinum nanoparticles inhibit antioxidant effects of vitamin C via ascorbate oxidase-mimetic activity. J Mater Chem B. 2016;4:7895–790.
100. Watanabe A, Kajita M, Kim J, Kanayama A, Takahashi K, Mashino T, Miyamoto Y. In vitro free radical scavenging activity of platinum nanoparticles. Nanotechnology. 2009;20 (45):455105.
101. Martín R, Menchón C, Apostolova N, Victor VM, Alvaro M, Herance JR, García H. Nano-jewels in biology. Gold and platinum on diamond nanoparticles as antioxidant systems against cellular oxidative stress. ACS Nano. 2010 Nov 23;4(11):6957–65.
102. Kajita M, Hikasaka K, Iitsuka M, Kanayama A, Toshima N, Miyamoto Y. Platinum nanoparticle is a useful scavenger of superoxide anion and hydrogen peroxide. Free Radic Res. 2007;41(6):615–26.
103. Rajkumar V, Guha G, Kumar RA. Antioxidant and anti-cancer potentials of Rheum emodi rhizome extracts. Evid Based Complement Alternat Med. 2011;2011:697986.
104. Ghosh S, Nitnavare R, Dewle A, Tomar GB, Chippalkatti R, More P, Kitture R, Kale S, Bellare J, Chopade BA. Novel platinum–palladium bimetallic nanoparticles synthesized by Dioscorea bulbifera: anticancer and antioxidant activities. Int J Nanomedicine. 2015;10:7477–90. https://doi.org/10.2147/IJN.S91579].
105. Shibuya S, Ozawa Y, Watanabe K, et al. Palladium and platinum nanoparticles attenuate aging-like skin atrophy via antioxidant activity in mice. PLoS One. 2014;9(10):e109288.
106. Kim J, Takahashi M, Shimizu T, et al. Effects of a potent antioxidant, platinum nanoparticle, on the lifespan of Caenorhabditis elegans. Mech Ageing Dev. 2008;129(6):322–31.
107. Szydłowska-Czerniak A. Łaszewska and A. Tułodziecka, a novel iron oxide nanoparticle-based method for the determination of the antioxidant capacity of rapeseed oils at various stages of the refining process. Anal Methods. 2015;7:4650–60.
108. Neupane BP, Chaudhary D, Paudel S, Timsina S, Chapagain B, Jamarkattel N, Tiwari BR. Himalayan honey loaded iron oxide nanoparticles: synthesis, characterization and study of antioxidant and antimicrobial activities. Internat J Nanomed. 2019;14:3533–41.
109. Narendhar Chandrasekar K, Kumar MM, Selvan K, Balasubramanian KK, Varadharajan R. Facile synthesis of Iron oxide, Iron-cobalt and zero valent Iron nanoparticles and evaluation of their antimicrobial activity, free radical Scavanging activity and antioxidant assay. Dig J Nanomat Biost. 2013;8(2):765–75.
110. Sastry NY, Padmaja JI, Rao RP, Kirani KRLS, Kaladhar DSVGK, Devi ST, Parvathi T, Gangadhar H, Kumar SK, Rao GD. In vitro dose dependent study on anti human pathogenic bacterial and free radical scavenging activities of methanolic seed coat extract of borassus flabellifer L. Asian J Pharm Clin Res. 2012;5:83–6.
111. Sandhya J, Kalaiselvam S. Biogenic synthesis of magnetic iron oxide nanoparticles using inedible borassus flabellifer seed coat: characterization, antimicrobial, antioxidant activity and in vitro cytotoxicity analysis. Mater Res Express. 2020;7:015045.
112. Shah STA, Yehya W, Saad O, Simarani K, Chowdhury Z, AAlhadi A, Al-Ani LA. Surface functionalization of iron oxide nanoparticles with gallic acid as potential antioxidant and antimicrobial agents. Nano. 2017;7:306.
113. Szekeres M, Illés E, Janko C, Farkas K, Tóth IY, Nesztor D, Zupkó I, Földesi I, Alexiou C, Tombácz E. Hemocompatibility and biomedical potential of poly (gallic acid) coated iron oxide nanoparticles for theranostic use. J Nanomed Nanotechnol. 2015;6:252.58.
114. Rasmussen JW, Martinez E, Louka P, Wingett DG. Zinc oxide nanoparticles for selective destruction of tumor cells and potential for drug delivery applications. Expert Opin Drug Deliv. 2010;7(9):1063–77.
115. Parkin G. Chemistry zinc–zinc bonds: a new frontier. Science. 2004;305:1117–8.

116. McCall KA, Huang C, Fierke CA. Function and mechanism of zinc Metalloenzymes. J Nutr. 2000;130:1437–46.
117. Stan M, Popa A, Toloman D, Silipas T-D, Vodnar DC. Antibacterial and antioxidant activities of ZnO nanoparticles synthesized using extracts of *Allium sativum*, *Rosmarinus officinalis* and *Ocimum basilicum*. Acta Metallurgica Sinica (English Letters). 2016;29(3):228–36.
118. Pavan Kumar MA, Suresh D, Nagabhushana H, et al. Beta vulgaris aided green synthesis of ZnO nanoparticles and their luminescence, photocatalytic and antioxidant properties. Eur Phys J Plus. 2015;130:109. https://doi.org/10.1140/epjp/i2015-15109-2.
119. Tasgin E. Green synthesis of zinc oxide (ZnO) nanoparticles and determination of it's antioxidant and antiradicalic activity. J Chromatogr Sep Tech. 2016;7(6(Suppl)) https://doi.org/10.4172/2157-7064.C1.020.
120. Shadmehri AA, Namvar F, Miri H, Yaghmaei P, Moghaddam MN. Assessment of antioxidant and antibacterial activities of zinc oxide nanoparticles, graphene and graphene decorated by zinc oxide nanoparticles. Int J Nano Dimens. Autumn 2019;10(4):350–8.
121. Ekezie FC, Suneetha J, Maheswari U. Biogenic synthesis of zinc nanoparticle from ethanol extract of bitter gourd and evaluation of its in-vitro antioxidant efficacy. IJISR. 2016;5:585–7.
122. Sanjay SS, Pandey AC, Kumar S, Pandey AK. Cell membrane protective efficacy of ZnO nanoparticles. Sop Trans Nano-Technol. 2014;1(1):21–9.
123. Pandey AC, Sanjay SS, Yadav RS. Application of ZnO nanoparticles in influencing the growth rate of Cicer Arietinum. J Exp Nanosci. 2010;5(6):488–97. *ISSN* 1745-8080 (Print), 1745-8099 (Online)
124. Sanjay SS, Pandey AC, Singh M, Prasad SM. Effects of functionalized ZnO nanoparticles on the phytohormones: growth and development of *Solanum melongena* L. *(*Brinjal*)* plant. World J Phar Res. 2015;4(5):1990–2009. ISSN: 2277-7105.
125. Stehbens WE. Oxidative stress, toxic hepatitis and antioxidants with particular emphasis on zinc. Exp Mol Pathol. 2003;75:265–76.
126. García-López JI, Zavala-García F, Olivares-Sáenz E, Lira-Saldívar RH, Díaz Barriga-Castro E, Ruiz-Torres NA, Ramos-Cortez E, Vázquez-Alvarado R, Niño-Medina G. Zinc oxide nanoparticles boosts phenolic compounds and antioxidant activity of *Capsicum annuum* L. during germination. Agronomy. 2018;8:215.
127. Arriagada F, Correa O, Günther G, Nonell S, Mura F, Olea-Azar C, Morales J. Morin flavonoid adsorbed on mesoporous silica, a novel antioxidant nanomaterial. PLoS One. 2016;11:e0164507.
128. Byszewski P, Klusek Z. Some properties of fullerenes and carbon nanotubes, SPIE proceedings series. Society of Photo-Optical Instrumentation Engineers; 2001.
129. Krusic PJ, Wasserman E, Keizer PN, et al. Radical reactions of C_{60}. Science. 1991;254:1183–5.
130. Gharbi N, Pressac M, Hadchouel M, Szwarc H, Wilson SR, Moussa F. Fullerene is a powerful antioxidant in vivo with no acute or subacute toxicity. Nano Lett. 2005;5(12):2578–85.
131. Slater TF, Cheeseman KH, Ingold KU. Carbon tetrachloride toxicity as a model for studying free-radical mediated liver injury. Philos Trans R Soc Lond Ser B Biol Sci. 1985;311:633–45.
132. Foley S, Crowley C, Smaihi M, et al. Cellular localisation of a water-soluble fullerene derivative. Biochem Biophys Res Commun. 2002;294:116–9.
133. Emelyantsev S, Prazdnova E, Chistyakov V, Alperovich I. Biological effects of C_{60} fullerene revealed with bacterial biosensor—toxic or rather antioxidant? Biosensors. 2019;9:81–91.
134. Kato S, Aoshima H, Saitoh Y, Miwa N. Bioorg Med Chem Letts. 2009;19(18):5293–6.
135. Xiao L, Takada H, Gan XH, et al. The water-soluble fullerene derivative 'radical sponge' exerts cytoprotective action against UVA irradiation but not visible-light-catalyzed cytotoxicity in human skin keratinocytes. Bioorg Med Chem Lett. 2006;16:1590–5.
136. Halliwell B. Reactive oxygen species and the central nervous system. J Neurochem. 1992;59:1609–23.
137. Sergeeva V, Kraevaya O, Ershova E, Kameneva L, Malinovskaya E, Dolgikh O, Konkova M, Voronov I, Zhilenkov A, Veiko N, Troshin P, Kutsev S, Kostyuk S. Antioxidant properties of

fullerene derivatives depend on their chemical structure: a study of two fullerene derivatives on HELFs. Oxidative Med Cell Longev. 2019Article ID 4398695;2019:13.
138. Kamat JP, Ghosh A, Devasagayam TPA. Mol Cell Biochem. 2000;209:47–53.
139. Beaudry F, Ross A, Lema PP, Vachon P. Phytother Res. 2010;24:525–30.
140. Kwon J, Kim J, Park S, Khang G, Kang PM, Lee D. Inflammation-responsive antioxidant nanoparticles based on a polymeric prodrug of vanillin. Biomacromolecules. 1618–1626;2013:14.
141. Del Nobile MA, Conte A, Incoronato AL, Panza O. J Food Eng. 2008;89(1):57–63.
142. Betbeder D, Lipka E, Howsam M, Carpentier R. Evolution of availability of curcumin inside poly-lactic-co-glycolic acid nanoparticles: impact on antioxidant and antinitrosant properties. Int J Nanomedicine. 2015;10:5355.
143. Das KC, Das CK. Curcumin (diferuloylmethane), a singlet oxygen ($^{(1)}O_{(2)}$) quencher. Biochem Biophys Res Commun. 2002;295(1):62–6.
144. Balasubramanyam M, Koteswari AA, Kumar RS, Monickaraj SF, Maheswari JU, Mohan V. Curcumin-induced inhibition of cellular reactive oxygen species generation: novel therapeutic implications. J Biosci. 2003;28(6):715–21.
145. Pu H-L, Chiang W-L, Maiti B, Liao Z-X, Ho Y-C, Shim MS, Chuang E-Y, Xia Y, Sung H-W. Nanoparticles with dual responses to oxidative stress and reduced pH for drug release and anti-inflammatory applications. ACS Nano. 2014;8:1213–21.
146. Chen X, Zou L-Q, Niu J, Liu W, Peng S-F, Liu C-M. The stability, sustained release and cellular antioxidant activity of curcumin Nanoliposomes. Molecules. 2015;20:14293–311.
147. Shah BR, Zhang C, Li Y, Li B. Bioaccessibility and antioxidant activity of curcumin after encapsulated by nano and Pickering emulsion based on chitosan-tripolyphosphate nanoparticles. Food Res Int. 2016;89:399–407.
148. Pool H, Quintanar D, de D. Figueroa J, Mano CM, Bechara JEH, Godínez LA, Mendoza S. J. Nanomater. 2012;2012. Article ID 145380, 12 pages
149. Wu TH, Yen FL, Lin LT, Tsai TR, Lin CC, Cham TM. Preparation, physicochemical characterization, and antioxidant effects of quercetin nanoparticles. Int J Pharm. 2008;346:160–8.
150. Casagrande R, Georgetti SR, Verri WA Jr, Dorta DJ, dos Santos AC, Fonseca MJ. Protective effect of topical formulations containing quercetin against UVB-induced oxidative stress in hairless mice. J Photochem Photobiol B. 2006;84:21–7.
151. Scalia S, Franceschinis E, Bertelli D, Iannuccelli V. Comparative evaluation of the effect of permeation enhances, lipid nanoparticles and colloidal silica on in vivo human skin penetration of quercetin. Skin Pharmacol Physiol. 2013;25:57–67.
152. Jain S, Jain AK, Pohekar M, Thanki K. Novel self-emulsifying formulation of quercetin for improved in vivo antioxidant potential: implications for drug-induced cardiotoxicity and nephrotoxicity. Free Radic Biol Med. 2013;65:117–30.
153. Carlotti ME, Sapino S, Ugazio E, Gallarate M, Morel S. Resveratrol in solid lipid nanoparticles. J Dispers Sci Technol. 2012;33:465–71.
154. Galati G, O'Brien PJ. Free Rad Bio and Med. 2004;37(3):287–303.
155. He C, Hu Y, et al. Effects of particle size and surface charge on cellular uptake and biodistribution of polymeric nanoparticles. Biomaterials. 2010;31(13):3657–66.
156. Charernsriwilaiwat N, Opanasopit P, Rojanarata T, Ngawhirunpat T. In vitro antioxidant activity of chitosan aqueous solution: effect of salt form. Trop J Pharm Res. April 2012;11 (2):235–42.
157. Zhai X, Zhang C, Zhao G, Stoll S, Ren F, Leng X. Antioxidant capacities of the selenium nanoparticles stabilized by chitosan. Nanobiotechnol. *2017*;15(*4*) https://doi.org/10.1186/s12951-016-0243-4.
158. Oh H, Lee JS, Sung D, Lee JH, Moh SH, Lim J-M, Choi WI. Synergistic antioxidant activity of size controllable chitosan-templated Prussian blue nanoparticle. Nanomedicine. Jul 2019;14 (19) https://doi.org/10.2217/nnm-2019-0223.

159. Nayak D, Minz AP, Ashe S, Rauta PR, Kumari M, Chopra P, Nayak B. Synergistic combination of antioxidants, silver nanoparticles and chitosan in a nanoparticle based formulation: characterization and cytotoxic effect on MCF-7 breast cancer cell lines. J Coll Interface Sci. 2016;470:142–52.
160. Anton N, Benoit JP, Saulnier P. Design and production of nanoparticles formulated from nano-emulsion templates – a review. J Control Release. 2008;128:186–96.
161. Sharma N, Bansal M, Visht S, et al. Nanoemulsion: a new concept of delivery system. Chron Young Sci. 2010;1:2–4.
162. Singh K, Ahmad Z, Shakya P, et al. Nano formulation: a novel approach for nose to brain drug delivery. J Chem Pharm Res. 2016;8:208–15.
163. Du L, Li J, Chen C, Liu Y. Nanocarrier: a potential tool for future antioxidant therapy. Free Radic Res. 2014 Sep;48(9):1061–9.
164. Rice-Evans CA. Measurement of total antioxidant activity as a marker of antioxidant status in vivo: procedures and limitations. Free Radic Res. 2000;33:S59–66.
165. Lindenmeier M, Burkon A, Somoza V. A novel method to measure both the reductive and the radical scavenging activity in a linoleic acid model system. Mol Nutr Food Res. 2007;51:1441–6.
166. Young IS. Measurement of total antioxidant capacity. J Clin Pathol. 2001;54:339.
167. Kirschbaum B. Total urine antioxidant capacity. Clin Chim Acta. 2001;305:167–73.
168. Lissi E, Salim-Hanna M, Pascual C, et al. Evaluation of total antioxidant potential (TRAP) and total antioxidant reactivity from luminol-enhanced chemiluminescence measurements. Free Radic Biol Med. 1995;18:153–8.
169. Van den Berg R, Haenen GRMM, van den Berg H, et al. Applicability of an improved Trolox equivalent antioxidant capacity (TEAC) assay for evaluation of antioxidant capacity measurements of mixtures. Food Chem. 1999;66:511–7.
170. Evelson P, Travacio M, Repetto M, et al. Evaluation of total reactive antioxidant potential (TRAP) of tissue homogenates and their cytosols. Arch Biochem Biophys. 2001;388:261–6.
171. Benzie IF, Strain JJ. Ferric reducing/antioxidant power assay: direct measure of total antioxidant activity of biological fluids and modified version for simultaneous measurement of total antioxidant power and ascorbic acid concentration. Methods Enzymol. 1999;299:15–27.
172. Cao G, Prior RL. Measurement of oxygen radical absorbance capacity in biological samples. Methods Enzymol. 1999;299:50–62.
173. Yan X, Murphy BT, Hammond GB, Vinson JA, Neto CC. Antioxidant activities and antitumor screening of extracts from cranberry fruit. J Agric Food Chem. 2002;50:5844–9.
174. Seeram N, Nair M. Inhibition of lipid peroxidation and structure-activity related studies of the dietary constituents anthocyanins, anthocyanidins and catechins. J Agric Food Chem. 2002;50:5308–12.
175. Vinson J, et al. Vitamins and especially flavonoids in common beverages are powerful in vitro antioxidants which enrich LDL and increase their oxidative resistance after ex vivo spiking in human plasma. J Agric Food Chem. 1999;47:2502–4.
176. Wolfe K, Liu RH. Cellular antioxidant activity (CAA) assay for assessing antioxidants, foods, and dietary supplements. J Agric Food Chem. 2007;55:8896–907.
177. Pisochi AM, Negulescu GP. Methods for total antioxidants activity determination. A. Review. Biochem Anal Biochem. 2011;1:1–10.
178. Pochinok TV, et al. A rapid method for determination of antioxidant activity of medicinal substances. Khim Farm Zh (in Russian). 1985;19(5):565–9.
179. P. Trogadas, J. Parrondo, F. Mijangos, V. Ramani (2011), ESI J. Mat. Chem. 16 pages.
180. Kappus H. Lipid peroxidation: mechanisms, analysis, enzymology and biological relevance. In: Sies H, editor. Oxidative stress. London: Academic Press; 1985. p. 273–310.
181. Lacatusu NB, Oprea O, Bojin D, Meghea A. J Nanopart Res. 2012;14:902.
182. Wettasinghe M, Shahidi IF. J Agric Food Chem. 1999;47(5):1801–12.
183. Glazer AN. Phycoerythrin fluorescence-based assay for reactive oxygen species. Methods Enzymol. 1990;186:161–8.

184. Cao G, Alessio HM, Cutler RG. Oxygen-radical absorbance capacity assay for antioxidants. Free Radic Biol Med. 1993;14:303–11.
185. Benzie IFF, Strain JJ. The ferric reducing ability of plasma (FRAP) as a measure of antioxidant power: the FRAP assay. Anal Biochem. 1996;239:70–6.
186. Kokubo K. Water-soluble single-Nano carbon particles: Fullerenol and its derivatives. London: InTech; 2012.
187. Lam CW, James JT, McCluskey R, Hunter RL. Toxicol Sci. 2004;77:126–34.
188. Soto KF, Carrasco A, Powell TG, Garza KM, Murr LE. J Nanopart Res. 2005;7:169–45.
189. Sayes CM, Fortner JD, Guo W, Lyon D, Boyd AM, Ausman KD, Tao YJ, Sitharaman B, Wilson LJ, Hughes JB, West JL, Colvin VL. Nano Lett. 2004;4:1881–7.
190. Sanjay SS. Precautions to avoid consequences leading to nanotoxification. In: Shukla AK, editor. ISBN 978-981-13-8953-5 ISBN 978-981-13-8954-2 (eBook) p. 201 Nanoparticles in medicine. Singapore: Springer Nature Singapore Pte Ltd.; 2020. p. 201–20.
191. Barriga-González G, Aguilera-Venegas B, Folch-Cano C, Pérez-Cruz F, Olea-Azar C. Electron spin resonance as a powerful tool for studying antioxidants and radicals. Curr Med Chem. 2013;20:4731. https://doi.org/10.2174/0929867311320990157.

Chapter 4
Mechanism of Antioxidant Activity

Abstract Basically the antioxidants activity follows mechanism based on the redox reactions. Redox reactions are very common in our day-to-day life. During the occurrence of various physiological reactions in our body, free radicals are also generated as inevitable by-products which may cause damage to our cells. Free radicals are generated through many pathways and promulgate through chain initiation to chain termination by the butting action of antioxidants. Neutralization of free radicals, ROS or RNS checks the oxidative damage and may operate in number of ways such as direct scavenging of free radicals, activating antioxidant enzymes, chelating metal catalysts, and inhibiting oxidases. At the molecular level, the operations are done through five known antioxidant mechanisms that can be used to describe the basis of various electron–proton transfer theories involving thermodynamic operators to describe antioxidant reactions.

Keywords Redox reactions · Free radical scavenging · Antioxidant mechanism · Nano-antioxidants · Nanocomposite

1 Introduction

The mechanism of antioxidants activity is based on the redox process, i.e., where oxidation and reduction process occurs simultaneously. Redox reactions are very common and important chemical reactions occurring in our day-to-day life. Many of the vital changes occurring in our biological system take place within our bodies and other living organisms through redox reactions only. As an example, in a process of digestion, carbohydrates, fat molecules, etc. present in our food breaks down to form carbon dioxide and water. In this process energy is released, which is utilized to perform our various daily activities. During the occurrence of these processes, free radicals are also generated as inevitable by-products which may cause damage to our cells through a series of reaction mechanisms.

S. Sundaram Sanjay, A. K. Shukla, *Potential Therapeutic Applications of Nano-antioxidants*, https://doi.org/10.1007/978-981-16-1143-8_4

1.1 Pathways for the Free Radical Production Activity

According to Harman's proposed "free radical theory," the damage to cellular macromolecules via free radical production in aerobic organisms is a major factor of life [1]. Free radicals are natural by-product of aerobic cell metabolism. They may be endogenous (within the body) or exogenous (outside the body) mentioned earlier. Endogenous free radical formation occurs continuously in the cells as a result of both enzymatic and non-enzymatic reactions. Endogenous free radicals are formed in the body via four different processes [2].

Process I In the first stage, as part of the natural metabolism of oxygen-dependent nutrients, energy molecules adenosine triphosphate (ATP) is released by mitochondria, the intracellular powerhouse, by consuming oxygen and produces water (aerobic respiration).

$$6O_2 + C_6H_{12}O_6 \rightarrow 6CO_2 + 6H_2O + \text{energy (ATP)}$$

$$\text{Main products} \rightarrow \text{carbon dioxide} + \text{water} + \text{energy}$$

$$\text{By} - \text{products} \rightarrow H_2O_2, OH^{\cdot}, \text{superoxide radicals etc.}$$

In this process, as a result of the oxygen molecule's incomplete reduction, hydrogen peroxide, hydroxyl and superoxide radicals, and other undesirable by-products are inexorably generated.

The superoxide radicals are primarily produced at two important locations in the electron transport chain during the metabolic process: complex I (NADH oxidoreductase/dehydrogenase) and complex III (NADH dehydrogenase) (cytochrome c oxidoreductase/reductase). The reduced form of coenzyme Q (QH2) is formed as a result of electron transfer from complex I or II to coenzyme Q or ubiquinone (Q). Reduced form QH2 regenerates coenzyme Q via an unstable intermediate semiquinone anion (Q-) in the Q-cycle. The intermediate Q- passes its electrons to molecular oxygen and forms the superoxide radical almost immediately. Since superoxide is not produced by enzymes, the triggering of ROS increases as the rate of metabolism rises [3]. The superoxide anion is then converted to hydrogen peroxide by mitochondrial superoxide dismutase. According to Starkov's observation, monoamino oxidase, α-ketoglutarate dehydrogenase, glycerol phosphate dehydrogenase, and p66shc are some of the other mitochondrial components that support and mediate the formation of ROS [4]. Generally p66Shc used to be found in cytoplasm, but also occurs in the mitochondrial inter-membrane space in fractions. Due to oxidative stress, p66Shc moves to the inter-membrane space of the mitochondria. Here it allies with cytochrome c and produces ROS generation [5]. During normal metabolism, approximately each cell produces above 20 billion oxidant molecules per day, resulting in dangerous free radicals as a by-product. It is very fact that oxygen, the elixir of life is the cause behind the human body's ailments, harmful and destructive effects.

Process II Oxidants (free radicals, ROS) such as nitric oxide, superoxide, hydrogen peroxide, and others are released by our immune system's fighters, the white blood cells, to destroy the attacking pathogenic microbes, parasites, bacteria, viruses, etc., as important component of a disease-fighting defense mechanism. Free radicals with unquenched electrons, on the other hand, cause a great mess by attacking nearby molecules in order to get electrons to attain stability. This attack destroys the victim molecules, resulting in the generation of a new free radical, which sets off a chain reaction. When the process is begun, it causes healthy living cells to be disrupted. As a result of the tears in the tissues, the system is damaged, wreaking havoc [6].

Process III Peroxisomes, an unintended consequence, a side-product formed due to the breakdown of the fatty acids and other cytosolic molecule degradation, produce hydrogen peroxide and heat, whereas in mitochondria fatty acids are oxidized to produce ATP and water. Usually an antioxidant enzyme, catalase, degrades these peroxides. However, some hydrogen peroxide spurts out and causes havoc in other parts of the cell. Since, fats with polyunsaturation have number of multiple double bonds, so, as they are oxidized, they produce an excess of free radicals derived from lipid peroxides [7].

Process IV One of the body's key defenses against harmful substances consumed with food is cytochrome P450, a cellular enzyme. It also causes production of oxidant by-products during fighting against foreign chemicals such as drugs and pesticides. In zest, free radicals strike virtually all of the body's organs and tissues constantly ripping our system like biological terrorists [8].

1.2 Promulgation Mechanism for Free Radical Productivity

Once the reaction is started, the mechanism for triggering of free radicals is promulgated in the following steps;

1. *Free radical chain initiation reaction:* The origination of free radicals in a reaction initiates chain reaction. This initiation produces reactive intermediates.
2. *Chain Propagation:* Another reactive intermediate is formed when the intermediate interacts with a stable product molecule. This becomes a cyclic process. As a result of which number of free radicals are formed.
3. *Chain Termination:* There may be several side reactions that eradicate the reactive intermediates, and as a result, the chain reaction gets terminated (Scheme 4.1).

Specific enzymes inside our bodies check these free radicals just after they are formed, termed as antioxidant enzymes (discussed earlier) or antioxidant nutrients found in our edibles, vitamins A, C and E, beta-carotene, flavonoid are some common examples. Free radicals generally play role of oxidants, and those that act as protective mechanisms against them are referred to as antioxidants, not reductants.

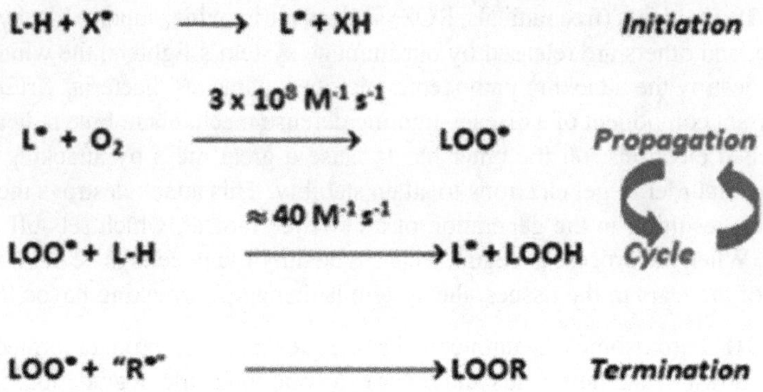

$$L\text{-}H + X^{\bullet} \longrightarrow L^{\bullet} + XH \quad \textit{Initiation}$$

$$L^{\bullet} + O_2 \xrightarrow{3 \times 10^8\ M^{-1}\,s^{-1}} LOO^{\bullet} \quad \textit{Propagation}$$

$$LOO^{\bullet} + L\text{-}H \xrightarrow{\approx 40\ M^{-1}\,s^{-1}} L^{\bullet} + LOOH \quad \textit{Cycle}$$

$$LOO^{\bullet} + \text{“}R^{\bullet}\text{”} \longrightarrow LOOR \quad \textit{Termination}$$

Scheme 4.1 Schematic representation of free radical formation and termination reactions

Antioxidants are the substances that protect tissues, cells, and organelles from free radical damage.

2 Antioxidant Activity

The cyclic chain propagation process can be terminated by some preventive measures that can break the chain reaction. Chain-breaking antioxidants slow down or stop oxidative processes by quenching the chain-carrying radicals. The assigned work of antioxidants is to check the free radicals to cause any damage to the body. Preventive antioxidants butts the oxidizing species before damage is done.

3 Mechanism of Antioxidant Activity

The mechanism through which antioxidants perform their function is basically by neutralizing or pairing of electron spins of oxidants that may be of free radicals, ROS or RNS. This neutralization checks the oxidative damage and may operate in number of ways, specifically by:

1. direct scavenging of free radicals,
2. scavenging and quenching of ROS and RNS,
3. activating antioxidant enzymes,
4. chelating metal catalysts,
5. sequestration of transition metal ions,
6. reducing alpha-tocopherol radicals,
7. inhibiting oxidases,
8. extenuating oxidative stress caused by nitric oxide,

9. ending of chain reactions by free radicals,
10. repairing molecules of radical's damages.

At the molecular level, all the abovementioned operations are done through five known antioxidant mechanisms that can be demonstrated using electron–proton transfer theories involving thermodynamic operators to describe antioxidant reactions [9–17]:

3.1 Hydrogen Atom Transfer (HAT) Mechanism

Antioxidants perform their free radical scavenging activity predominantly by providing electrons or hydrogen to them. In consonance with this mechanism, an antioxidant interacts with a free radical directly, and neutralizes free electron of the free radicals by donating hydrogen to them. Then there occurs formation of a complex between, say, the lipid radical and the antioxidant radical (free radical acceptor). As a result, antioxidants get oxidized (Scheme 4.2), and free radical gets neutralized.

Proper antioxidants perform their antioxidant behavior by making lipid free radicals or phenolic compounds inactive. As with phenolic compounds, hydroperoxide equalizers prevent hydroperoxides from crumbling into free radicals. Synergistic effects promote the activities of conventional antioxidants, as in case of citric acid and ascorbic acid. Here, one should be very cautious while selecting an antioxidant (A) and should take care of the following preventive measures:

1. Both antioxidant and antioxidant free radical ($A^{\bullet}$) should be relatively unreactive.
2. $A^{\bullet}$ should decay to give products that are safe.

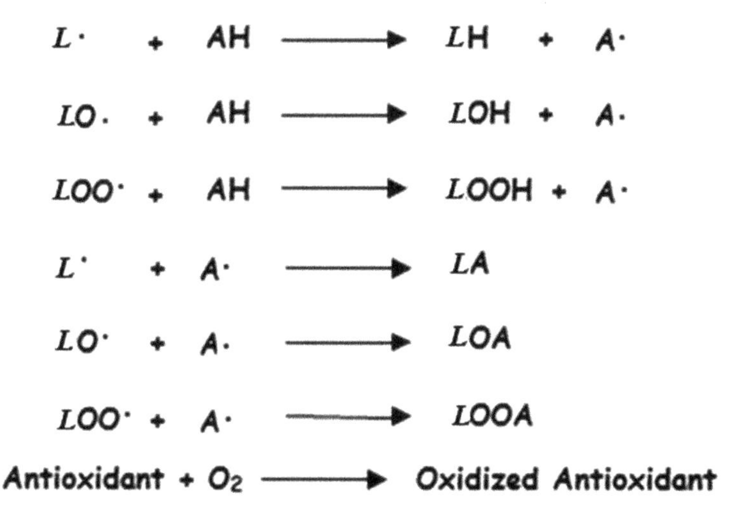

Scheme 4.2 Reactions of antioxidants with radicals

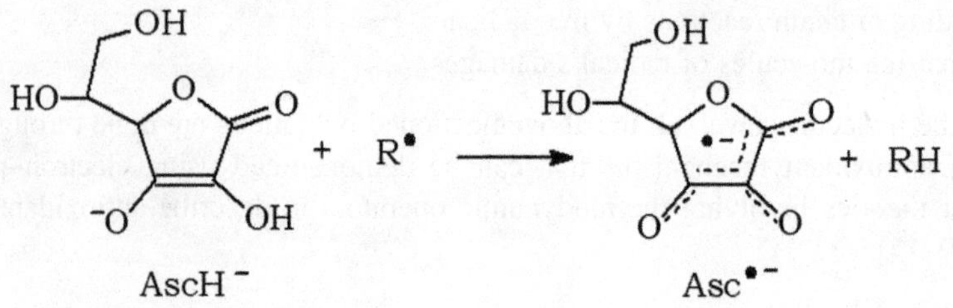

$AscH^- + R^\bullet \longrightarrow Asc^{\bullet -} + RH$

Scheme 4.3 Ascorbic acid providing hydrogen to free radicals

3. To form a peroxyl radical, it shouldn't add any oxygen.
4. It can be renewable or recycled.
5. If the antioxidant breaks the chain by donating hydrogen, it ought to be at the top of the **hierarchy** in food chain.

As an example, ascorbic acid is a very good hydrogen donor. Ascorbic anion ($AscH^-$) serves as a contributor antioxidant [18], donating a hydrogen atom having single electron ($H^\bullet$) or electron to a proton ($H^+ + e^-$) forming resonance-stabilized tricarbonyl ascorbate free radical which is generated by an oxidizing radical (Scheme 4.3). Since pKa is −0.86 for AscH•, therefore, it is present as Asc•- and it has not been protonated in the biological system.

Another natural donor antioxidant, tocopherol works as

$$LOO^\bullet + TOH \rightarrow LOOH + TO^\bullet$$

Peroxide radicals ($ROO^\bullet$) are generally chain carriers. Arresting R• may also break the chain reaction. Though it is uncommon in biological systems, in the polymer sector, it plays a pivotal role.

$$RH + ROO^\cdot \rightarrow ROOH + R^\cdot$$
$$R^\cdot + O_2 \rightarrow ROO^\cdot$$

In case of RNS, nitric oxide also serves in the capacity of a sacrificial antioxidant as

$$LOO^\bullet + {}^\bullet NO \rightarrow LOONO$$

$^\bullet NO$ can terminate chain reaction by reacting with oxy-radicals. In this way, $^\bullet NO$ upregulates the systems that contribute to the antioxidant network.

$$ROO^\cdot + {}^\cdot NO \rightarrow ROONO$$
$$RO^\cdot + {}^\cdot NO \rightarrow RONO$$

Phenolic compounds make up a large number of antioxidants, which use resonance structures to stabilize their radical forms after receiving hydrogen, which allows the unpaired electron to be attached to an oxygen atom in which the unpaired electron can be placed on an oxygen atom or the nearby aromatic ring's para carbon [19]. Due to sp2-hybridized carbon lattice, graphene is highly aromatic but isn't a very committed H-donor. As a result, it is unlikely to have any antioxidant activity. Graphenes, on the other hand, may behave as a pro-oxidant. Owing to different structures, graphene oxide or graphene with a few layers reveal ROS inhibition and antioxidant activity [20].

Again, now let us see the antioxidant activity in the view of thermodynamic considerations:

The O–H links' bond dissociation energy (BDE) is a significant criterion for assessing antioxidant activity. If O-H bond weakens, the release of hydrogen atoms is facilitated and easier will be the reaction of free radical inactivation.

$$\mathrm{LOH} + \mathrm{X}^{\cdot} \rightarrow \mathrm{LO}^{\cdot} + \mathrm{XH} \tag{4.1}$$

Here L represents more often an aryl group/antioxidant.

The numerical parameter associated with this mechanism is "*Bond Dissociation Enthalpy*" (BDE), which may be calculated with the help of following equation:

$$\mathrm{BDE} = \mathrm{H}_{\mathrm{LO}^{\cdot}} + \mathrm{H}_{\mathrm{H}^{\cdot}} - \mathrm{H}_{\mathrm{LOH}} \tag{4.2}$$

in which $\mathrm{H}_{\mathrm{LO}^{\cdot}}$, $\mathrm{H}_{\mathrm{H}^{\cdot}}$, and $\mathrm{H}_{\mathrm{LOH}}$ are the enthalpy's of the radical, H-atom, and of the reduced compound, respectively. Better antioxidant properties are associated with a species having lesser BDE value.

3.2 *The SET: Single Electron Transfer Mechanism*

In this case, a single electron from a free radical is transferred to phenolic antioxidant. The antioxidant may offer a free radical an electron, and it becomes a positively charged radical. Lower ionization potential (IP) value makes easier abstraction of electron:

$$\mathrm{LOH} + \mathrm{X}^{\cdot} \rightarrow \mathrm{LOH}^{\cdot+} + \mathrm{X}^{-} \tag{4.3}$$

The SET mechanism has a numeral variable called "*Adiabatic Ionization Potential*" (AIP), that may be evaluated with the help of following equation:

$$\mathrm{AIP} = \mathrm{H}_{\mathrm{LOH}^{\cdot+}} - \mathrm{H}_{\mathrm{LOH}} \tag{4.4}$$

here $H_{LOH\cdot+}$ and H_{LOH} are the enthalpies of cationic radical and of the compound, respectively.

3.3 SET-PT: Single Electron Transfer Accompanied with Proton Transfer

$$LOH + X^{\cdot} \rightarrow LOH^{\cdot+} + X^{-} \quad (4.5)$$

$$LOH^{\cdot+} \rightarrow LO^{\cdot} + H^{+} \quad (4.6)$$

A two-step reaction is involved in this mechanism. The free radical reacts with a phenolic antioxidant molecule in the first step, where the phenolic antioxidant appears as a cationic radical form and other one as an anionic radical form. This reaction is a crucial step in this two-step process in terms of thermodynamics. The phenolic antioxidant's cationic radical breaks up into a phenolic radical and a proton in the second step. Adiabatic Ionization Potential (AIP) for the first step and Proton Dissociation Enthalpy (PDE) for the second step are numeral variables associated with the SET-PT process, which can be given as:

$$PDE = H_{LO}^{\cdot} + H_{H}^{+} - H_{LOH}^{\cdot+} \quad (4.7)$$

in which $H_{LO}^{\cdot}$, H_{H}^{+}, and $H_{LOH}^{\cdot+}$ are the enthalpies of the radical, the proton, and of cationic radical.

3.4 SPLET: Sequential Proton Loss Electron Transfer

$$LOH \rightarrow LO^{-} + H^{+} \quad (4.8)$$

$$LO^{-} + X^{\cdot} + H^{+} \rightarrow LO^{\cdot} + XH \quad (4.9)$$

A two-step reaction is present in this mechanism as well. The phenolic antioxidant splits into an anionic form and a proton in the first step. The free radical then reacts with the ions formed in the first reaction. As a result of which, a phenolic antioxidant radical and a neutral molecule pop up in this reaction. The numeral variable for these processes are as follows: "*Proton affinity*" (PA) is for the first reaction step and "*Electron Transfer Enthalpy*" (ETE) is for the second step that can be given as Eqs. (4.10) and (4.11):

Scheme 4.4 Antioxidants-mechanisms representations by equations: (1) HAT, (2) SET, (3) &(4) SET-PT, (5) & (6) SPLET

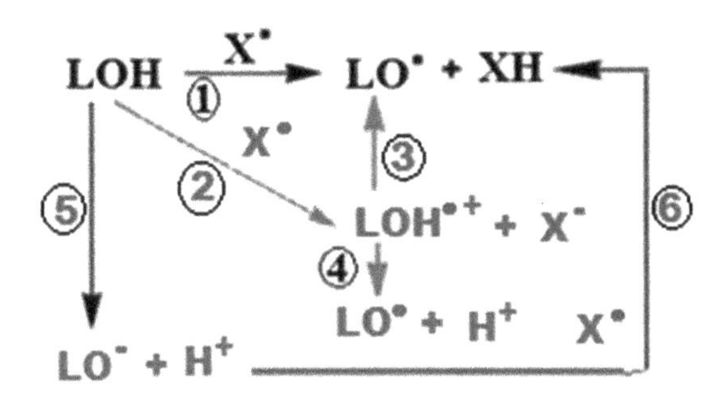

$$PA = H_{LO}{}^{-} + H_{H}{}^{+} - H_{LOH} \tag{4.10}$$

here $H_{LO}{}^{-}$, H_{H+}, and H_{LOH} are the enthalpies of the anion, proton, and of the compound, respectively.

$$ETE = H_{LO}{}^{\cdot} - H_{LO}{}^{-} \tag{4.11}$$

here $H_{LO}{}^{\cdot}$ and H_{LO-} are the enthalpies of the radical and of the anion, respectively.

Equations (4.1)–(4.9) are represented in following Born–Haber type cycle showing consumption of free radicals ($X^{\cdot}$) in number of steps (Scheme 4.4).

3.5 *Transition Metals Chelation (TMC)*

Ferritin, lactoferrin, albumin, and ceruloplasmin are few metal binding proteins that bind free iron (Fe) and copper (Cu) ions, which have capacity to catalyze oxidative reactions.

Iron and copper are the principal metals which are mainly targeted to produce free radicals because they are normally loosely bound to proteins. Transferrin, hemoglobin, and ceruloplasmin are the examples of some biomolecules in which there exist protein-metal bonding. There are many chelating agents which may form strong chelates with metal ions. For example, Fe^{3+} forms chelates with EDTA, DTPA, Deferrals, etc. and Fe^{2+} with phenanthrolines, salicylic acids, etc. Iron in hemes, as well as loosely bound iron on proteins and DNA, may be toxic. Iron and oxygen reactions are an important route for the chain initiation for free radical formations in biological oxidations, because they promote hazardous oxidant production as following reactions (4.12)–(4.15) [21]:

$$\text{Chelates of Fe(II)/Cu(I)} + H_2O_2 \rightarrow HO^{\bullet} + \text{chelates of Fe(III)/Cu(II)} + OH \tag{4.12}$$

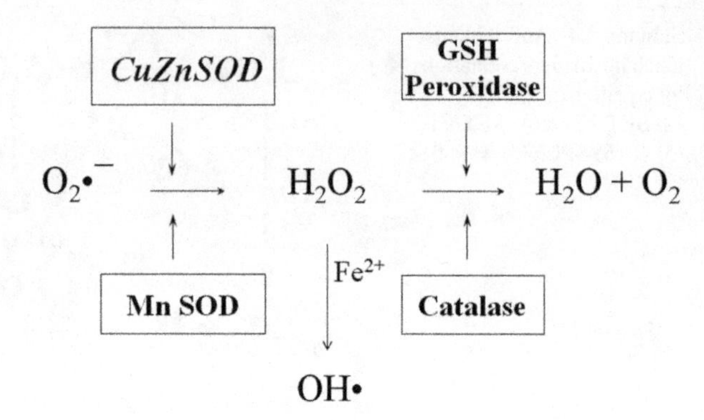

Fig. 4.1 Metal-based oxygen radical defense enzymes

$$\text{Chelates of Fe(II)/Cu(I)} + \text{LOOH} \rightarrow \text{LO}^\bullet + \text{chelates of Fe(III)/Cu(II)} + \text{LOH} \quad (4.13)$$

$$\text{Chelates of Fe(II)/Cu(I)} + O_2 \rightarrow \text{Oxidants} \quad (4.14)$$

Preventive measures can be taken with $^\bullet$NO radical, in case of Fe(II)chelate. It forms a coordinate compound with heme-iron,

$$\text{heme-Fe}^{2+} + {}^\cdot\text{NO} \rightarrow \text{heme-Fe}^{2+} - \text{NO} \quad (4.15)$$

The role of metal-based oxygen radical defense enzymes can be shown pictorially as Fig. 4.1:

Further, phosphoric acid, Millard reaction compounds, citric acid, etc. are few examples of metal chelators that bind heavy metals to form inactive compounds. Quenchers of singlet oxygen present in carotenes perform its antioxidant activity by converting singlet oxygen (^{1}O) into triplet oxygen (^{3}O). Hydroperoxides are reduced by substances like proteins and amino acids in a non-radical approach, in which these act upon as reducing hydroperoxides [22].

The antioxidant mechanism in metals occurs where metals with lower oxidation states, in accordance with Fenton reaction, produce free radicals [23]:

$$H_2O_2 + M^{m+} \rightarrow OH^- + OH^\cdot + M^{(m+1)+} \quad (4.16)$$

$$\text{LOH} \rightarrow \text{LO}^- + H^+ \quad (4.17)$$

Any dissociable organic molecule can form chelate with metals. Polyphenol anions, in particular, have a strong tendency to form chelated complex with metals. Since deprotonated hydroxyls in polyphenols are often involved in metal chelation, a molecule's capability to generate the proton has to be taken into account. The numeral variable associated with this mechanism is referred to as "*gas phase acidity*," which corresponds to the compound's vacuum enthalpy, $\Delta H_{acidity}$.

$$\Delta H_{acidity} = H_{LO}{}^{-} - H_{LOH} \tag{4.18}$$

in which $H_{LO}{}^{-}$ and H_{LOH} are the enthalpies of the anion and of the compound, respectively.

The free Gibbs energy ($\Delta G_{acidity}$) is measured if the assessment is done in solvents.

$$\Delta G_{acidity} = G_{LO}{}^{-} - G_{LOH} \tag{4.19}$$

here $G_{LO}{}^{-}$ and G_{LOH} are the Gibbs free energy of the anion and of the compound, respectively.

Distribution of electrons in HOMO and LUMO, as well as the spin density of molecules, are other significant descriptors of antioxidant properties. A molecule's ability to donate a proton is impaired due to its lower HOMO capacity. It means that the HOMO distribution identifies the chemical groups in a molecule that are vulnerable to free radical attack. The spin density of a molecule describes the non-paired electron distribution and the stability of a molecule's radical conformation. We can deduce the degree of chemical activity of a molecule from the difference in LUMO and HOMO energy levels. The following formula was used to calculate the energy difference ΔE between LUMO and HOMO energy levels:

$$\Delta E = E_{LUMO} - E_{HOMO} \tag{4.20}$$

$\Delta E_{(LUMO - HOMO)}$ is inversely proportional to molecule's activity. Its low value is indicative of molecule's lower reactivity [24].

4 Mechanism for the Functioning of Nano-Antioxidants

Since the nature and behavior of nanoparticles depend on its structural morphology, size, shape, capping or functionalized groups, their mechanism of functioning as nano-antioxidants also varies accordingly. Some of the most probable theories are discussed below to explain its antioxidant mechanism.

4.1 Electron-Hole Excitonic Pairs ($e-,H^{+}$)Theory

We know that at nanoscale level, the behavior of electron can be explained only by quantum mechanics and not by Newtonian classical mechanics. Therefore, in case of nano-sized metal oxides, there occurs a unique property (quantum mechanical property) of electron-hole excitonic pairs ($e-,h+$) in their conduction and valence band in significant numbers, which are accessible to institute redox reactions. The

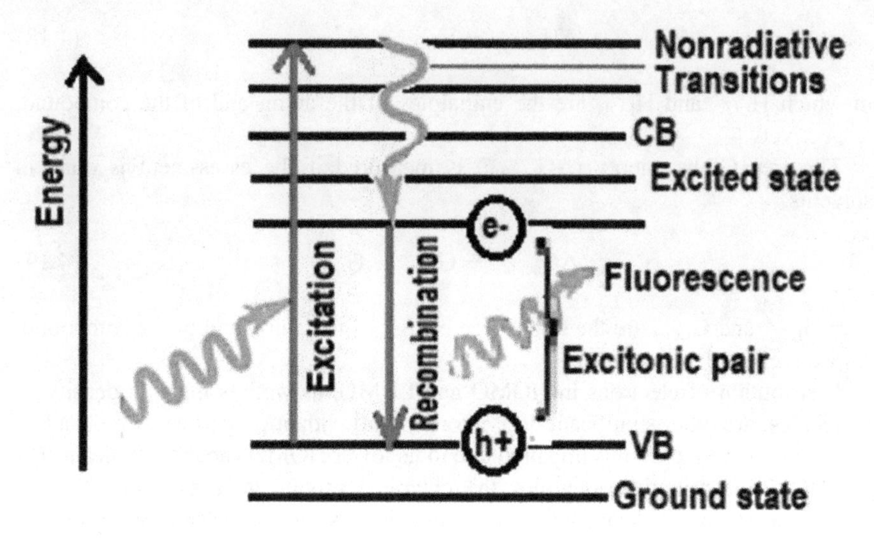

Fig. 4.2 Excitonic pair (e^-,h^+) in conduction (CB) and valence band (VB)

electrons in the excited state are extremely energetic, and they combine with free radicals almost instantly with free radicals thus preventing the recombination of e−, h+ pair. Metal oxide nanoparticles can protect cells against the harmful effects of oxidative stress by capturing reactive oxygen species or free radicals. Consequently, measuring oxidative stress criterion, i.e., assaying oxidative damage products, provides significant information and better understanding about the pathophysiology of several diseases. Secondly, nanoparticles have free hanging bonds at the surface. This causes flaws in its crystalline form, that are much more prominent here than in bulk materials. Moreover, as a result of the increased disorder in the lattice structure, more metal ions lying at the interstitial space of the lattice dissolve and interact with cell organelles.

In doing so, the number of oxygen vacancies also increases, thus serving as potential antioxidants. A photoluminescence analysis of ZnO NPs revealed the existence of oxygen vacancies which also contributes to structural disorders [25] (Fig. 4.2).

Elswaifi and coauthors also supported this theory while studying the behavior of cerium oxide nanoparticles (CeONP). They mentioned that, at the atomic level, the behavior of CeO nanoparticles is governed by quantum mechanical properties. As CeONP has a high potential for electron absorption and transfer, Elswaifi and colleagues came to the conclusion that these nanoparticles could also act as catalysts for absorbing and transferring atomic bodies. This catalytic action, dispensed by CeONP, reinstates the cell's redox balance in an indirect manner, as it allows electrons to flow more easily and can be expressed on the quantum scale [26].

Chistyakov et al. have explained the antioxidant action of fullerenes with the help of quantum theory and taking into account of the documented evidence that fullerene accumulates primarily in mitochondria [27]. They hypothesized that to achieve antioxidant action of fullerenes, only a small reduction in mitochondrial potential is needed. Being negatively charged, the mitochondrial inner membrane located in the respiratory system may serve as a source of harmful reactive oxygen species (ROS). The outer mitochondrial membrane is positively charged. If fullerenes could gain a positive charge by consuming an excess of protons (may be a h^+), it would be a huge breakthrough there, they'd be able to carry positive charge through the mitochondrial membrane, reducing mitochondrial potential. This mechanism would gracefully explain and give reasons for the slight support of the fullerenes in decoupling of respiration and phosphorylation. Furthermore, fullerenes considered as carbon nanotubes equivalents are expected to have akin mechanism, since they were also found to have mitochondrial congregations and to have an antioxidant effect [28–30].

4.2 π-π Stacking Interactions Theory

Mechanism for the nano-antioxidants can be deduced to some level from their structural characteristics. The addition of phenolic groups to capped AuNPs could improve the reaction kinetics of nano-antioxidants like PEG3SA capped Au. Libo Du and his co-scientists [31] juxtaposed the spectral variations between PEG3SA capped Au and salvianic acid. When compared to salvianic acid, they found that PEG3SA capped Au displayed a significant redshift (from 267 to 278 nm) in its overall maximum absorption.

The redshift suggested the π-π piling up interactions amidst the vicinal phenolic groups bedaubed on the surface of AuNPs. Piling up of π-π interactions describes the attractive, non-covalent interlinkages that occur in allying aromatic rings. Such π-π interactions is responsible for the increased rate of the self-assembled nano-antioxidants reaction. This concept may help to design more efficient nano-antioxidants. Thus the kinetics of the antioxidants reflects the occurrence of antioxidant activity, which is primarily mediated by the free radicals based on oxygen and other species of reactive oxygen (ROS). Nie et al. observed that the direct involvement of metal nanoparticles in scavenging the DPPH radical is responsible for the increased antioxidant efficacy of nano-antioxidants [32].

4.3 Electron Abstraction Theory (EAT)

Libo Du's [31] group have found that in living cells, the antioxidant-functionalized AuNPs' reinforced kinetic effect in ROS scavenging is sustained. Inside a living cell, they observed that oxygen-centered free radicals produced by tBuOOH can easily withdraw an electron from polyunsaturated fatty acid to produce a carbon-based lipid radical within a living cell.

This produces a lipid peroxyl radical as it interacts with molecular oxygen. If endogenous antioxidants would not be able to reduce the resulting peroxyl radicals in a flash, lipid peroxidation will occur as a result of the triggering of free radical chain reaction. **Warheit** found that the three-carbon dialdehyde malonyldialdehyde (MDA) is formed as a by-product of polyunsaturated lipid peroxidation and it is commonly used as a biomarker to assess lipid peroxidation levels [33].

As per the findings of Wu and his co-workers, the antioxidative property of ZnO nanoparticles may be attributable to the shifting of electron density from the oxygen atom to the odd electron found on the nitrogen atom of DPPH. The thermal stability of nanoparticles is determined by the available free energy of oxides, which is dependent on the structural configuration of oxygen atoms [34].

Xiaoli and his team discovered microperoxidase's (MP) direct electron transfer with the aid of ZnO nanoparticles, when the oxide's effect on microperoxidase was studied electrochemically. Initially, MP was not prepared to conduct redox reactions on an electrode surface but the protein showed direct quick electron transfer with the aid of ZnO NPs [35]. Their studies revealed that the immobilization of MP with ZnO NPs encourages the catalytic activity on the electrode surface against H_2O_2. They observed that the addition of H_2O_2 into the test solution showed different responses on the cathodic peak of MP and anodic peak of the enzyme with gradual increase of the H_2O_2 concentration. This is due to MP's catalytic reduction of H_2O_2 in the presence of nano-ZnO.

4.4 *Hydrogen Abstraction Theory (HAT)*

There are number of mechanisms that contribute to the fullerenol nanoparticles' (FNP) antioxidant action. Djordjevic's group administered the hydroxyl radical's ability to extract a hydrogen from fullerenol, as well as the origination of a fullerenol radical, $C_{60}(OH)_{23}O\cdot$, that is rather more stable [36]. Fullerenol nanomolecules form hydrogen bonds with H_2O and other fullerenol molecules in aqueous solution, resulting in negatively charged anionic nanoparticles that are very stable. A hydroxyl radical can also remove one electron from fullerenol and a radical cation, $C_{60}(OH)_{24}^{+}$ is resulted.

Husebo et al. have suggested a mechanism, considering the fact that the poly-anionic nanoparticles have several free electron pairs from oxygen that are spread around the FNP and have a high potential for the formation of covalent coordinate bond with metal ions that are pro-oxidant [37].

Mirkov et al. demonstrated that FNP inhibits lipid peroxidation in a liposome-based cell membrane model system [38]. DMPO-spin trap/ESR method was used by Kokubo to assess the antioxidant potential of the water-soluble fullerene $C_{60}(OH)_{32}\cdot 8H_2O$.

The peaks in ESR spectrum assigned to the hydroxyl radicals that were reduced by this fullerene derivative. At the same time, a singlet radical-signal was not due to OH increase on increasing $C_{60}(OH)_{32}\cdot 8H_2O$ concentrations. The implication of these findings is that the dehydrogenation of $C_{60}(OH)_{32} \cdot 8H_2O$ causes it to scavenge $OH^{\cdot}$ and is oxidized to a stable fullerenol radical at the same moment [39].

As discussed much that the execution of scavenging effect is an important mechanism of antioxidant activity. Applying DPPH type of steady free radicals, this effect can be easily measured by means of EPR spectroscopy. Since DPPH has a nitrogen-centered free radical that is relatively stable and to achieve an even diamagnetic configuration, the molecule readily absorbs an electron or a hydrogen radical with a great ease from the substance capable of donating electrons or hydrogen atoms.

The fading of the intensity of purple colored DPPH mixed solution has been observed due to the formation of 1, 1-diphenyl-2-picrylhydrazine (yellow) (non-radical form of DPPH), is dependent on the number of electrons taken up from the donor results in the reduction in the EPR signal intensity [41].

4.5 *Electrostatic Attraction Theory*

In the case of ZnO NPs obtained via plant extracts assisted synthetic process, the electrostatic attractive interaction between negatively charged bioactive compounds of plant extracts and positively charged nanoparticles is possibly responsible for the antioxidant potency against DPPH. In this driving force that the phytochemicals bind to ZnO NPs, and enhances their activity in the biological systems in a synergistic manner [41].

The variations in scavenging behavior of ZnO prepared with various plant extracts were discovered to be a function of nanoparticle's size and consequently diverse specific surface area exposed for the scavenging activity. The maximum radical scavenging activity was found in ZnO samples with the smallest nanoparticle size. Au nanoparticles showed a similar consistent pattern. Due to their lower specific surface area, negligible scavenging behavior was observed in a solution of the nanocomposite containing larger sized Au nanoparticles [42, 43].

References

1. Harman MD. Aging: a theory based on free radical and radiation chemistry. J Gerontology. 1956;11(3):298–300.
2. Lobo V, et al. Free radicals, antioxidants and functional foods: impact on human health. Pharmacogn Rev. 2010;4(8):118–26.
3. Finkel T, Holbrook NJ. Oxidants, oxidative stress and the biology of ageing. Nature. 2000;408:239–47.
4. Starkov AA. The role of mitochondria in reactive oxygen species metabolism and signaling. Ann N Y Acad Sci. 2008;1147:37–52.

5. Giorgio M, Migliaccio E, Orsini F, Paolucci D, Moroni M, Contursi C, et al. Electron transfer between cytochrome c and p66Shc generates reactive oxygen species that trigger mitochondrial apoptosis. Cell. 2005;122(2):221–33.
6. Dhalla NS, et al. Role of oxidation stress in cardiovascular diseases. Hypertension. 2000;18 (6):655.
7. Balaban RS, Nemoto S, Finkel T. Mitochondria, oxidants, and aging. Cell. 2005;120:483–95.
8. Colbert MD. What you Don't know may be killing you. Siloam Pub; 2000.
9. Szeląg M, Mikulski D, Molski M. Quantum – chemical investigation of the structure and the antioxidant properties of α-lipoic acid and its metabolites. J Mol Mod. 2011; https://doi.org/10.1007/s00894-011-1306-y.
10. Leopoldini M, Russo N, Toscano M. The molecular basis of working mechanism of natural polyphenolic antioxidants. Food Chem. 2011;125:288–306.
11. Foti MC, Daquino C, Geraci C. Electron – transfer reaction of cinnamic acids and their methyl esters with the DPPH. Radical in alcoholic solutions. J Org Chem. 2004;69:2309–14.
12. Litwinienko G, Ingold KU. Abnormal solvent effects on hydrogen atom abstraction. 2. Resolution of the curcumin antioxidant controversy. The role of sequential proton loss electron transfer. J Org Chem. 2004;69:5888–96.
13. Musialik M, Litwinienko G. Scavenging of dpph. Radicals by vitamin E is accelerated by its partial ionization: the role of sequential proton loss electron transfer. Org Lett. 2005;7:4951–4.
14. Nakanishi I, Kawashima T, Ohkubo K, Kanazawa H, Inami K, Mochizuki M, Fukuhara K, Okuda H, Ozawa T, Itoh S, Fukuzumi S, Ikota N. Electron – transfer mechanism in radical scavenging reactions by vitamin E model in a protic medium. Org & Biomolecular Chem. 2005;3:626–9.
15. Litwinienko G, Ingold KU. Solvent effects on the rates and mechanisms of reaction of phenols with free radicals. Acc Chem Res. 2007;40:222–30.
16. Leopoldini M, Russo N, Toscano M. Gas and liquid phase acidity of natural antioxidants. J Agric Food Chem. 2006;54:3078–85.
17. Mikulski D, Szeląg M, Molski M, Górniak R. Quantum chemical study on the antioxidation mechanisms of transresveratrol reactions with free radicals in the gas phase, water and ethanol environment. J Mol Structure. 2010;951:37–48.
18. Buettner GR, Jurkiewicz BA. The ascorbate free radical as a marker of oxidative stress: an EPR study. Free Radic Biol Med. 1993;14:49–55.
19. Everett SA, Dennis MF, Patel KB, Maddix S, Kundu SC, Willson RL. Scavenging of nitrogen dioxide, Thiyl, and sulfonyl free radicals by the nutritional antioxidant -carotene. J Biol Chem. 1996;271:3988–94.
20. Qiu Y, Wang Z, Owens ACE, Kulaots I, Chen Y, Kane AB. Antioxidant chemistry of graphene-based materials and its role in oxidation protection technology. Nanoscale. 2014;6:11744–55.
21. Qian SY, Buettner GR. Iron and dioxygen chemistry is an important route to initiation of biological free radical oxidations: an electron paramagnetic resonance spin trapping study. Free Radic Biol Med. 1999;26:1447–56.
22. Kumar S. Asian J Res in Chem Pharm Sc. 2014;1(1):27–44.
23. Fenton HJH. Oxidation of tartaric acid in the presence of iron. J Chem Soc Trans. 1894;65:899–910.
24. Sebastian S, Sundaraganesan N, Manoharan S. Molecular structure, spectroscopic studies and first-order molecular hyperpolarizabilities of ferulic acid by density functional study. Spectrochim Acta A Mol Biomol Spectrosc. 2009;74:312–23.
25. Sanjay SS, Pandey AC, Kumar S, Pandey AK. Cell membrane protective efficacy of ZnO nanoparticles. Sop Transactions Nano-Technol. 2014;1(1):21–9.
26. Elswaifi SF, Palmieri JR, Hockey KS, Rzigalinski BA. Antioxidant nanoparticles for control of infectious disease. Infect Disord Drug Targets. 2009;9:445–52.
27. Chistyakov VA, Smirnova YO, Prazdnova EV, Soldatov AV. Possible mechanisms of fullerene C_{60} antioxidant action. Bio Med Res Int. 2013;2013:821498.

28. Chistyakov VA, Zolotukhin PV, Prazdnova EV, Alperovich I, Soldatov AV. Physical consequences of the mitochondrial targeting of single-walled carbon nanotubes probed computationally. Physica E Low DimensSystNanostruct. 2015;70:198–202.
29. Zhou F, Xing D, Wu B, Wu S, Ou Z, Chen WR. New insights of transmembranal mechanism and subcellular localization of noncovalently modified single-walled carbon nanotubes. Nano Lett. 2010;10(5):1677–81.
30. Ma X, Zhang LH, Wang LR, et al. Single-walled carbon nanotubes alter cytochrome c electron transfer and modulate mitochondrial function. ACS Nano. 2012;6(12):10486–96.
31. Du L, Suo S, Wang G, Jia H, Liu KJ, Zhao B, Liu Y. Chem Eur J. 2013;19:1281–7.
32. Nie Z, Liu KJ, Zhong CJ, Wang LF, Yang Y, Tian Q, Liu Y. Enhanced radical scavenging activity by antioxidant-functionalized gold nanoparticles: a novel inspiration for development of new artificial antioxidants. Free Radic Biol Med. 2007;43:1243–54.
33. Warheit DB. Nanoparticles: Health risks?. Mater Today. 2004:32–35.
34. Wu TH, Yen FL, Lin LT, Tsai TR, Lin CC, Cham TM. Preparation, physicochemical characterization and antioxidant effects of quercetin nanoparticles. Int J Phar. 2008;4:346(1-2):160–8.
35. Zhu X, Yuri I, Gan X, Suzuki I, Li G. Electrochemical study of the effect of nano-zinc oxide on microperoxidase and its application to more sensitive hydrogen peroxide biosensor preparation. Biosens Bioelectron. 2007;22:1600–4.
36. Djordjevic A, Canadanovic-Brunet JM, Vojinovic-Miloradov M, Bogdanovic G. Oxid Commun. 2004;27(4):806–12.
37. Husebo LO, Sitharaman B, Furukawa K, Kato T, Wilson LJ. J Am Chem Soc. 2004;126 (38):12055–64.
38. Mirkov SM, Djordjevic AN, Andric NL, et al. Nitric Oxide Bio Chem. 2004;11(2):201–7.
39. Kokubo K. Water-soluble single-Nano carbon particles: Fullerenol and its derivatives. London: InTech; 2012.
40. Morsy MA. Spectrosc Int J. 2002;16:371.
41. Kumar B, Smita K, Cumbal L, Debut A. Bioinorg Chem Appl. 2014;2014:523869.
42. Yakimovich NO, Ezhevskii AA, Guseinov DV, Smirnova LA, Gracheva TA, Klychkov KS. Russ Chem Bull. 2008;57:520.
43. Watanabe A, Kajita M, Kim J, Kanayama A, Takahashi K, Maino T, Miyamoto Y. Nanotechnology. 2009;20:1.

Chapter 5
Quantification of Antioxidants

Abstract With the advent of new and improved computational methods, quantitative structure-activity relationship, QSAR, is one of such exploring fields getting a passionate attention. QSAR is a useful tool for elucidating the mechanisms of antioxidant action of these specific nanoparticles. It helps in designing unique and innovative nanoparticles which have the capability to be used as therapeutics. QSAR equations create a connection between a collection of molecules' various molecular properties and their experimentally established biological active molecular descriptive. The activities are assessed by chemical measurements and their biological assays. Precise logical representations of a molecule are known as molecular descriptors that are generated as a result of a well-defined algorithm, and then extended to a well-defined molecular template or an analytical process. Many descriptors can be focused entirely on quantum mechanical calculations directly or could also be calculated using the probability distribution of molecule's electronic wave function. Generally quantitative structure-property/activity relationships (QSPRs/QSARs) are estimators, but they play a crucial role in the assessment of molecular structures. For converting the dataset into training and test set, QSAR/QSPR modelling normally involves four main operations, viz., calculation, selection, generation, and validation. For data analysis, mostly all QSPR modelling techniques use some sort of logistic regression equations. In studying the polyphenols, they are a perfect entity for QSAR analyses due to their extensive substitution patterns. There is not a single method that may be considered to work best for all the problems. The nature of a substance may differ from other in any respect. Thus QSPR/QSAR studies are based on traditional descriptors or on specific nano-descriptors depending on the material or nanomaterial.

Keywords Quantitative structure-property/activity relationships (QSPRs/QSARs) · Bioactivity · Molecular descriptors · QSAR/QSPR modelling · Phytochemicals

S. Sundaram Sanjay, A. K. Shukla, *Potential Therapeutic Applications of Nano-antioxidants*, https://doi.org/10.1007/978-981-16-1143-8_5

1 Introduction

Today is the era of computational experiments for quick results. With the advent of newer and better analytical techniques, as well as fast computer methodological advancements the work to get results in certain specific direction and framing of the research sketches accordingly become amenable. One of these analytical domains is QSAR, or quantitative structure-activity relationship that is drawing a lot of interest. Sensor molecules have been discovered to be excellent evokers for the activation of plant secondary metabolites such as polyphenols. Signal constituents have recently been used as an effective tool for producing desired supplemental bioactive metabolites in plant tissues.

However, it is currently uncommon in professional drives. As a signalling component, for the diverse metabolic and physiological reactions, nanomaterials are being used and becoming an interesting ground for the recent researchers. QSAR is a useful tool for elucidating the significance of these particular nanoparticles' antioxidative activity. These will also assist in the designing of innovative and more effective nanoparticles, which have the potency that could be used as chemotherapeutic drugs. The QSAR model assumes that there exist an essential correlation between molecular structural system and a specific bioactivity. Thus accordingly QSAR equations make efforts to find the connection between various molecular properties and experimentally recognized bioactivities of a group of molecules. That is, the basic idea behind QSAR is to use chemical impressions to transform compounds into those with required features and proficiency into a digitalized and mathematically assessed form.

2 What is QSAR?

QSAR/QSPR/QPAR is a precise mathematical conceptual based model which is obtained from a group of molecules with known behavioral properties, pursuits, toxicity, and so on, and evaluated using computational methods. These quantitative relationships occur when the property and/or activity are a quantitative one. It is not necessary that all properties and activities of chemical compounds can be quantitized. We have to observe and decide according to the requirement and results. We don't have any unanimous accepted quantitatively expressed properties that have same level of accuracy which can be expressed.

According to Crum-Brown and Fraser [1], *"performing upon a substance a chemical operation which shall introduce a known change into its constitution, and then examining and comparing the physiological action of the substance before and after the change,"* it is possible to create a relation between chemical composition and physiological activity. Such quantitative relationships between these two are studied under the paradigm of QSAR (quantitative structure-activity relationship), QSPR (quantitative structure-property relationship), and QPAR (quantitative property-activity relationship). QSAR, QSPR, and QPAR are often abbreviated as (Q)SAR, (Q)SPR, and (Q)PAR, or simply SAR, SPR, and PAR. QSAR studies provide a method to identify the existence of specific structure of compounds by the type of atoms or bonds present in it through electronic transitions and/or molecular vibrations and/or rotations or by the change in any other physicochemical properties that support in quantitative manner. It may be a problem of simple "affirmation" or "negation."

QSAR has now been extensively used in simulation of drug designing and research methodologies. Several researchers are using this approach to design relatively new antioxidant molecules with increased activity [2]. QSAR methodology helps in gaining the study on insight into structural features related to the various compounds to develop drugs having enhanced curative activities such as anticancer, antioxidant, and antifungal activities.

3 Molecular Descriptors

QSAR/QSPR represents an attempt to associate the physicochemical or structural properties of a group of structurally similar molecules with the support of some physical properties or physiological behavior that may be related to therapeutic, toxicological, or environmental factors. The electrical structures and properties, hydrophobicity, wettability, topological effects, and steric effects are generally accounted for by molecular descriptors. Chemical experiment and biological assays are used to evaluate their activities. Getting relevant data for identifying descriptors for a specific molecule or its fragmented segment is a crucial component for progressing QSAR technique. When it is done, then, with the help of QSARs data predictive models of biological activity can be drawn and which may lead to the mechanism of activity.

The mathematical representations of a molecule are known as molecular descriptors that are generated using a well-defined algorithm and implemented on an explicit molecular expression or a research process. They provide the means through which the chemical details that are encrypted within a molecular structure can be represented numerically using a mathematical approach. More clearly, molecular descriptors are the ultimate outcome of a logical and mathematical process that converts chemical details incorporated within a logical consequences of a molecule into a useful number or the outcome of a standard prescribed experiment. Basic molecular descriptors can be developed by counting particular atom types or structural segments in the molecule, as well as physicochemical and bulk properties, including total of primary, secondary, tertiary, quaternary carbon or, secondary, tertiary or quaternary carbons in ring, substituted or non-substituted aromatic carbon, unsaturation index, hydrophilic factor, molecular refractivity, the number of hydrogen bond donors or acceptors, the number of OH-groups, and so forth. Algorithms using topological description may yield additional molecular descriptors, commonly called as topological, or two dimensional descriptors. The three-dimensional spatial

Table 5.1 List of some descriptors

Class of descriptors	Descriptor type
Topological	E-state parameters, connectivity indices, kappa shape indices
Structural	H-bond acceptor, H-bond donor, number of OH-groups, Rotlbonds, Chiral centers
Physicochemical	Molecular weight, pressure, concentration, pH, solubility
Thermodynamic	Critical temperature, logarithmic partition coefficient (LogP), Van der Waals force, molar refractivity, melting point, boiling points
Spatial(Steric Parameters)	Indices of shadow, gyrational radius, surface area of a molecule, denseness, moment of inertia Volume of a molecule, Van der waals' radii, Van der waals' volume, molar refractivity, etc.
Electronic	Dipole moment, energy of configurations, ionization constant, pKa, chemical shifts in IR and NMR
Magnetic	Paramagnetic, diamagnetic, ferro or ferromagnetic, antiferromagnetic

(x, y, z) coordinates of the molecule can be used to generate other molecular descriptors often known as geometrical or three-dimensional descriptors. Here, statistical correlation approaches are used to investigate the quantitative relationship between a group of compounds' biological activities and their attributes. Four-dimensional descriptors deduced from molecule-to-molecule interactional energies and some probe, which are deeply ingrained in a grid. The capability to describe each molecule by an assemblage of geometrical conformations, orientations at various alignments and protonation states is the fourth dimension in 4D QSAR. It greatly decreases the biases associated with ligand's orientation and coordination site selection. Fifth dimension in 5D QSAR and thus 5D descriptors creates the ability to display up to six separate inflicted models as an ensemble. The 6D-QSAR method allows several solvation systems to be evaluated at the same time. For these higher versions of QSAR studies, software programs are also available [3].

Alternatively, innumerable descriptors can be determined precisely from quantum-mechanical computations or from the molecules' electronic wave function or electrostatic field. Because the electrophilicity coefficient is a chemical reactivity based descriptor whose formulation is dependent on the density functional theory [4, 5], these descriptors are entirely suitable to be used in our QSAR paradigm. The electrophilicity coefficient has recently been employed as a viable descriptor of bioactivities, revealing that it accurately assesses biological functions. Some quantum chemical descriptors, including soft texture, chemical potential, and electrophilicity quotient, may be employed just within the chassis of density functional theory if they display excellent co-relation in the estimation of radical inhibiting antioxidant capacity [6] (Table 5.1).

Thus, for developing a QSAR model, there are a variety of molecular descriptors accessible. Of course, most molecular descriptors are probably irrelevant in assessing biological functions. Therefore, finding the optimal grouping of descriptors plays a pivotal role in deciding the activity. The data can be analyzed using an appropriate statistical method combined with an adjustable shortlisting process to create a QSAR model only with the group of descriptors that are far more statistically

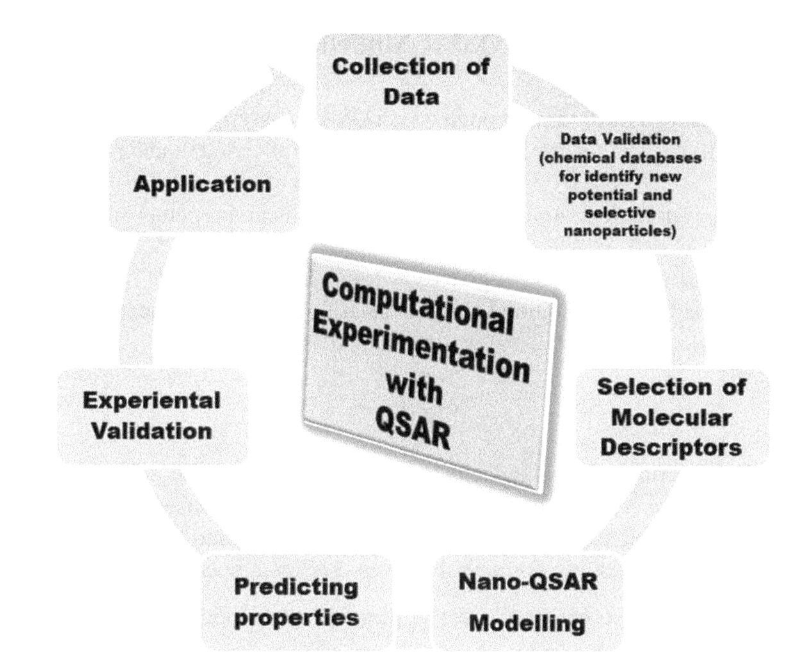

Fig. 5.1 Route for the QSAR operations

important for determining bioactivity. Traditional quantitative structure-property/activity relationships (QSPRs/QSARs) are approximations but it has a significant impact on molecular structure prediction [7–11]. Of course, it is not an easy job. It is also affected by a variety of other factors, such as physicochemical processing conditions, the purity of the drug, and so on. For studying different biological activities, the nature of organisms, cell types, and other factors also put an ostensible impact. Subsequently, different outputs must also be weighed in light of their interdependence. First is based on the molecule's structure and second, that's via surrounding positions.

We know that the molecular geometry depends on the environment on which molecule exists such as temperature, pressure, boiling points, pH, concentration, thermodynamic parameters and accordingly contributes therapeutic effects or toxicity. In Structure-Property-Activity Relationships (SPARs) approach, a natural selection algorithm is developed for obtaining "Molecular Descriptors Structure-Activity Relationship" for given sets of compounds and given property or activity such as molecular topology, molecular geometry, relativities (as biological activities) and its evidences (as physicochemical properties) [12] (Fig. 5.1).

4 Operations for QSAR/QSPR Modelling

The general mathematical form through which QSAR analysis for the evaluation of a biological activity is done can be represented by the following equation.

$$\text{Biological Activity} = f\ (\text{Physicochemical Property})$$

Crum-Brown and Fraser have given the following equation in 1868, which is perhaps the first QSAR definition [1]. As a linear function of the chemical structure *C*, the "physiological activity" (Φ) was calculated as:

$$\Phi = f_{(C)}$$

For the validation of a model, the required dataset must be split into two: a training dataset (for creating the QSAR model) and a testing dataset to investigate its consistency for biological activities. To convert the dataset into training and testing set, QSAR/QSPR modelling normally involves four main operations:

1. calculation or measurement of a set of descriptors or another type of reference parameters,
2. selection of a limited fraction of these descriptors that are significant only to the biological activity under consideration,
3. generation of the descriptors' nonlinear connectivity to the extensive attributes of the substance,
4. validation of the model's dependability, consistency, accuracy, and universal adaptability scope [13].

Regression analysis is done in most of the QSPR modelling approaches. The regression equations for the data analysis are generally developed by using: (1) stepwise regression, with simple least squares, (2) multiple linear regression was used after factor analysis (FA-MLR) together with (3) assessment using partial least squares analytical process (PLA).

4.1 Progression Through Stepwise Regression

In stepwise progression [14] Step by step, a multiple component linear equation is constructed, (a) an initial design must be identified in the first stage, (b) second is the iteration or reduplication stepping process. Here changing of the preceding step's model is done by inserting or eliminating dependent variables according to the scaling or stepping criteria, for example, among the values obtained for certain parameter, say R = 5 inserted and R = 4.9 eliminated, (c) finally, when stepping is no longer possible according to the stepping requirements, or if a given highest possible number of measures is achieved, the search is concluded. For every step, all variables are examined and weighed to choose which one will make the largest contribution in the equation. The process is then restarted after that variable is incorporated in the model. This is how the iteration process proceeds.

4.2 Factor Analysis Followed by Multiple Linear Regression (FA-MLR) (Linear Regression with Many More Variables)

In FA-MLR scenarios where the structure-property correlation is nonlinear, a polynomial, bilinear, or neural net, in such circumstances of FA-MLR, an established approach to the MLR technique is employed as the ultimate statistical tool for constructing conventional QSAR relationships [15, 16]. In the data-preprocessing step, FA is applied to detect the key descriptors that contribute to the predictor variables and to duck collinearities between them. The data matrix is first normalized in such a classic FA process, followed by the construction of the minimized covariance matrices. The component sequence can then be retrieved from the relevant eigen-vectors by solving an eigen-value problem.

4.3 Partial Least Squares Analysis (PLSA)

PLSA is a type of regression that can deal with data that is highly correlated, noisy, or has a large number of X variables [17, 18]. It produces a simplified solution that is statistically more reliable than MLR. The PLS linear model identifies "new variables" which are comprised of linear combinations of the predictor dataset. To prevent overfitting, a stringent evaluation for the importance of each successive PLS factor is required, followed by a closure when another elements are no longer important. PLS enables for the creation of larger QSAR equations while avoiding overfitting and deleting the majority of possible factors. To acquire the optimal number of components, PLS is frequently used in conjunction with cross examination. This assures that the QSAR equations are designed for their capacity to identify it, instead of just fitting data. MLR is the simplest QSPR modelling method. It is considered that the descriptors are a good approximation of the property being simulated [19].

5 QSAR for Phytochemicals

In case of phytochemicals such as polyphenols or flavonoids, they have a phenyl-benzo-γ-pyrone structure in common (Scheme 5.1).

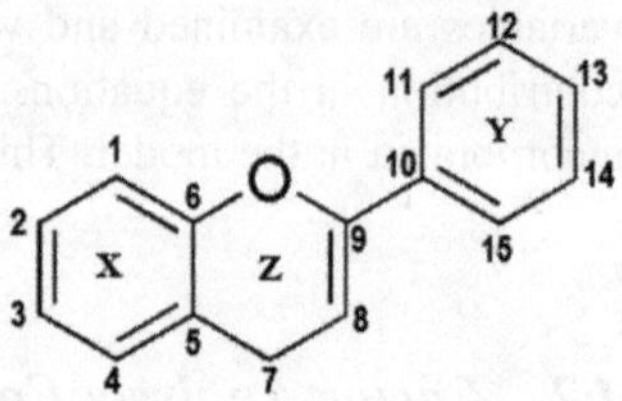

Scheme 5.1 The common flavonoids framework

Among the thousands of flavonoids, for their antioxidant properties the differentiation between them is not an easy job. Only the location and numbers of hydroxyl and/or methoxyl groups, including the placement and number of distinct saccharides involved in glycosylation, distinguish them from one another (Fig. 5.2). So the property of acylation, which frequently occurs at various locations of the flavonoid core and glycosyl moieties, may be exploited for the evaluation of their antioxidant properties. Because of the numerous substitution sequences, flavonoids or polyphenols are suitable units for QSAR researches [6].

There is not a single method that may be considered to work best for all the problems. The goal of statistical simulation is to create correlation models amongst independent predictor variables (molecular descriptors) and dependent responsive variables (biological properties). Simple linear regression analysis, multivariate linear regression, principal component analysis, partial least squares regression, genomic function approximation, and genomic partial least squares approaches are some of the methods that are accessible [20, 21].

6 QSAR Modelling Patterns for Nanomaterials

A range of nanoparticles, as well as many other agents, have been found effective and exciting materials in this area over the last decade [22]. The nature of a substance may differ from other in any respect. It may be its toxicity, mutagenicity, or its carcinogenicity. QSPR/QSAR studies are based on traditional descriptors or on specific nano-descriptors as mentioned above, depending upon the concern nanomaterials.

The multiple activities of nanomaterials and their structural diversity make these compounds an opulent source for the modelling to study their pharmacological properties. The activity of nanoparticles as mentioned earlier is due to its enhanced surface to volume ratio and the presence of its specific discrete electronic energy levels. Its surface can be functionalized with different groups to make them lipid or water soluble so that they become physiologically active.

In most of the cases, the use of generalized strategies are preferably expected for both conventional QSPR/QSAR and predictive models, based on preparation methods, that are focused to nanomaterials, Although, this perfect condition is almost certainly impossible to achieve. Therefore, the development of models for

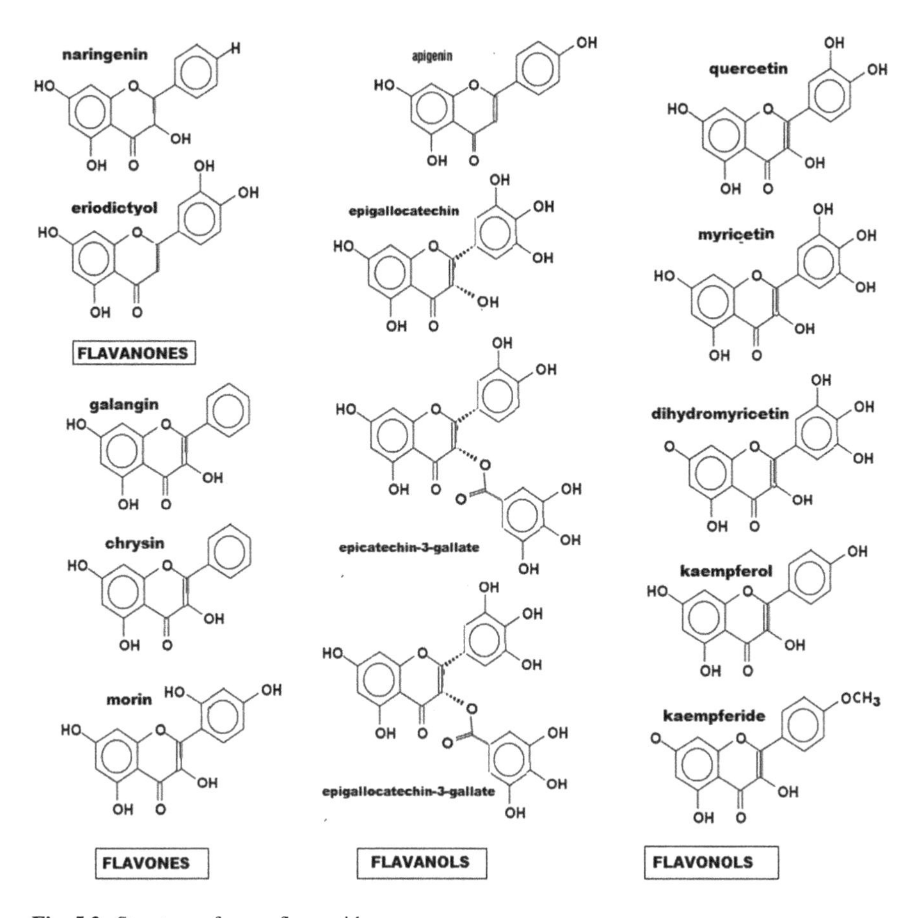

Fig. 5.2 Structure of some flavonoids

conventional entities and the progression of models for nanomaterials has to be established individually in the coming years.

With the help of nano-QSAR/nano-QSPR modelling we can predict the properties necessary to understand the mechanisms of actions of nanoparticles before experimentation and thus ease the expenditures of experiments. This makes a consistent bonding between structural, morphological, and physicochemical properties of nanoparticles to weigh the biological activity.

To develop membrane interaction quantitative structure-activity relationship (MI-QSAR) models for understanding the interactions of a carbon nanotube with a completely hydrated dimyristoyl-phosphatidylcholine (DMPC) lipid bilayer, Liu and Hopfinger have used molecular dynamics simulation in order to investigate the associations with both peritoneal permeating and phospholipid monolayer assemblies in their physiologically important liquid-crystalline process [23]. Thus for in vivo experimentation MI-QSAR methodology may serve a mode to access the modalities in studying the interaction of nanoparticles and nanotubes with the structural organization and dynamic attributes of lipid bilayers, transportation of minuscule polar molecules through these bilayers both with and without nanomaterial. Then the method to ascertain how far the passage of solvated ions via the incorporated nanotube affects the membrane bilayer's structure is determined. Solvated calcium ions are observed to have more weightage to be transported via the incorporated nanotube as compared to hydrated sodium ions. However, as the solvated calcium ions move in through the nanotube, they may bring about changes in the bound bilayer structure. Liu and Hopfinger have reported that, in the presence of the nanotube, the diffusion of the penetrants occurs quite effectively via the lipid bilayers, thus providing a means for the identification and prioritization of molecular properties for the plausible evaluation of toxicity or antioxidant properties of nanomaterials.

For reflecting the structural and behavioral heterogeneity of nanomaterials, while doing nano-QSAR both theoretically and experimentally derived descriptors, and the solutions adopted for modelling should not be the same as used for classical one [24]. Usually the computational methods are very powerful tool for dealing with nanoscale particles starting from initial quantum mechanical techniques. These methods permit to correlate the structural and electronic properties of nanoparticles in a logical way. Physicochemical properties such as size, agglomerated form, size distribution and morphology, porosity, lattice parameters, surface area, interfacial chemistry, charge density, chemical composition, and structure-dependent electronic configuration can be related to the antioxidant properties of nanoparticles [25]. Moreover, computational methods may reliably predict physico-molecular properties such as dipole moment, molecular volume, molecular electrostatic potentials, transition energies, ionization potentials, electron affinities, energy of valence orbitals, Fermi energy levels, band difference, and electronic charges. In the QSPR/QSAR methodologies, these computational properties could be used as molecular descriptors for defining structure property/activity relationships analysis.

Prior selection of suitable physicochemical molecular descriptors becomes a major restraint on working with nano-QSAR/nano-QSPR modelling because it is very challenging to express the specific antioxidant capability of nanoparticles, and then relating it with precise biological activities. This could be done only after performing a relative comparison between the experimental and computational results. A conventional in vitro nano-assays are generally easier to probe biological effects as inflammatory responses, oxidative stress, etc. to study antioxidant activities. But usually cell viability is targeted in computational studies. Moreover, for computational modelling, complicated mathematical equations have to be framed to predict a biological end point accurately to access human health risk evaluations.

7 Pros and Cons of QSAR Modelling

Advantages

1. The effect of structure on activity can be determined by quantifying the correlation between structure and its behavioral action.
2. QSAR studies provide a layout of newer synthetic route that may give improved activity.
3. The results could be very helpful in understanding the interactions between functional groups in the molecules that have the highest level of interaction with the biological earmarks [26].

Disadvantages

1. Due to determinate and indeterminate experimental errors, the quality of biological data may not be appropriate. This may give rise to false correlations to the considerable extent.
2. For QSAR analysis, a large number of training dataset (molecular descriptors' data) are required to reflect complete property evaluation. Lack of required amount of dataset, QSAR results cannot be used assuredly to predict the compounds of the precise activity.
3. For reliable results, 3D geometrical skeletal structures displaying ligand-receptor binding is needed. But in many a case 3D skeletons are not available. So the calculated results may not reflect characteristic features and may not be reliable as well.

References

1. Crum-Brown A, Fraser TR. On the connection between chemical constitution and physiological action. Pt 1. On the physiological action of the salts of the ammonium bases, derived from Strychnia, Brucia, Thebia, Codeia, Morphia, and Nicotia. T Roy Soc Edin. 1868-1869;25:151–203.
2. Rastija V, Medic-Saric M. QSAR study of antioxidant activity of wine polyphenols. Eur J Med Chem. 2009;44:400–8. https://doi.org/10.1016/j.ejmech.2008.03.001.
3. Puzyn T, Leszczynski J, Cronin MTD. Recent advances in QSAR studies. New York: Springer; 2010. p. 30–41.
4. Parr RG, Yang W. Chemical potential derivatives. In: Density-functional theory of atoms and molecules. 1st ed. New York: Oxford University Press; 1989. p. 87–95.
5. Parthasarathi R, Subramanian V, Royb DR, Chattarajb PK. Electrophilicity index as a possible descriptor of biological activity. Bioorg Med Chem. 2004;12:5533–43.
6. Hall LH, Mohney B, Kier LB. The electrotopological state: an atom index for QSAR. Quant Struct-Act Rel. 1991;10:43–51.
7. Pasha FA, Cho SJ, Beg Y, Tripathi YB. Quantum chemical QSAR study of flavones and their radical-scavenging activity. Med Chem Res. 2008;16:408–17.
8. Quesada-Romero L, Caballero J. Docking and quantitative structure-activity relationship of oxadiazole derivates as inhibitors of GSK3 \upbeta β. Mol Divers. 2014;18(1):149–59.
9. Achary PGR. QSPR modelling of dielectric constants of π-conjugated organic compounds by means of the CORAL software. SAR and QSAR Environ Res. 2014;25(6):507–26.

10. Achary PGR. Simplified molecular input line entry system-based optimal descriptors: QSAR modelling for voltage-gated potassium channel subunit Kv7.2. SAR and QSAR Environ Res. 2014;25:73–90.
11. Worachartcheewan A, Nantasenamat C, Isarankura-Na-Ayudhya C, Prachayasittikul V. QSAR study of H1N1 neuraminidase inhibitors from influenza a virus. Lett Drug Des Discov. 2014;11 (4):420–7.
12. Debnath AK. Quantitative structure-activity relationship (QSAR): a versatile tool in drug design. In: Ghose AK, Viswanadhan VN, editors. Combinatorial library design and evaluation: principles, software, tools and application in drug discovery. New York: Marcel Dekker Inc; 2001. p. 73–129.
13. Singh S, Supuran CT. 3D-QSAR CoMFA studies on sulfonamide inhibitors of the Rv3588c β-carbonic anhydrase from Mycobacterium tuberculosis and design of not yet synthesized new molecules. J Enzyme Inhib Med Chem. 2014;29(3):449–55.
14. Darlington RB. Regression and linear models. New York: McGraw Hill; 1990.
15. Franke R. Theoretical drug design methods. Amsterdam: Elsevier; 1984.
16. Franke R, Gruska A. Principal component and factor analysis. In: Van de Waterbeemd H, editor. Chemometric methods in molecular design (Methods and principles in medicinal chemistry), Vol. 2, R. Manhold, P. Krogsgaard-Larsen and H. Timmerman (Eds). Weinheim: VCH; 1995. p. 113–26.
17. Wold S. PLS for multivariate linear modeling. In: Van de Waterbeemd H, editor. Chemometric methods in molecular design (Methods and principles in medicinal chemistry). Weinheim: VCH; 1995. p. 195–218.
18. Fan Y, Shi LM, Kohn KW, Pommier Y, Weinstein JN. Quantitative structure-antitumor activity relationships of camptothecinanalogs: cluster analysis and genetic algorithm-based studies. J Med Chem. 2001;44:3254–63.
19. Le T, Epa VC, Burden FR, Winkler DA. Quantitative structure–property relationship modeling of diverse materials properties. Chem Rev. 2012;112:2889–919.
20. Tropsha A. Best practices for QSAR model development, validation and exploitation. Mol Inf. 2010;29:476–88.
21. Rusyn, et al. Predictive modeling of chemical hazard by integrating numerical descriptors of chemical structures and short-term toxicity assay data. Tox Sci. 2012;127(1):1–9.
22. Panneerselvam S, Choi S. Nanoinformatics: emerging databases and available tools. Int J Mol Sci. 2014;15(5):7158–82.
23. Liu J, Hopfinger AJ. Chem Res Toxicol. 2008;21:459–66.
24. Ahmadi S, Ghanbari H, Lotfi S, et al. Predictive QSAR modeling for the antioxidant activity of natural compounds derivatives based on Monte Carlo method. Mol Divers. 2020;25:87–97.
25. Oberdörster G, Maynard A, Donaldson K, Castranova V, Fitzpatrick J, Ausman K, Carter J, Karn B, Kreyling W, Lai D, Olin S, Monteiro-Riviere N, Warheit D, Yang H. Principles for characterizing the potential human health effects from exposure to nanomaterials: elements of a screening strategy. Part Fibre Toxicol. 2005;2:8.
26. Rasulev B, Gajewicz A, Puzyn T, Leszczynska D, Leszczynski J. Chapter 10, Nano-QSAR: advances and challenges, rsc nanoscience & nanotechnology No. 25; towards efficient designing of safe nanomaterials: innovative merge of computational approaches and experimental techniques. In: Puzyn T, Leszczynski J, editors. The royal society of chemistry. Published by the Royal Society of Chemistry; 2012. www.rsc.org.

CPSIA information can be obtained
at www.ICGtesting.com
Printed in the USA
LVHW081647030921
696883LV00002B/101